Studies in Big Data

Volume 167

The series "Studies in Big Data" (SBD) publishes new developments and advances in the various areas of Big Data- quickly and with a high quality. The intent is to cover the theory, research, development, and applications of Big Data, as embedded in the fields of engineering, computer science, physics, economics and life sciences. The books of the series refer to the analysis and understanding of large, complex, and/or distributed data sets generated from recent digital sources coming from sensors or other physical instruments as well as simulations, crowd sourcing, social networks or other internet transactions, such as emails or video click streams and other. The series contains monographs, lecture notes and edited volumes in Big Data spanning the areas of computational intelligence including neural networks, evolutionary computation, soft computing, fuzzy systems, as well as artificial intelligence, data mining, modern statistics and Operations research, as well as self-organizing systems. Of particular value to both the contributors and the readership are the short publication timeframe and the world-wide distribution, which enable both wide and rapid dissemination of research output.

The books of this series are reviewed in a single blind peer review process.

Indexed by SCOPUS, EI Compendex, SCIMAGO and zbMATH.

All books published in the series are submitted for consideration in Web of Science.

Marius Iulian Mihailescu · Stefania Loredana Nita ·
Valentina Marascu · Valentin Barna

Applied Quantum Computing and Cryptography

Challenges, Opportunities, and Performance Analysis for Algorithms

Springer

Marius Iulian Mihailescu
Faculty of Engineering and Computer Science
University of Spiru Haret
Bucharest, Romania

Stefania Loredana Nita
Faculty of Information Systems and Cyber Security
Military Technical Academy "Ferdinand I"
Bucharest, Romania

Valentina Marascu
Plasma and Radiation
National Institute for Laser, Plasma and Radiation Physics (INFLPR)
Măgurele, Romania

Faculty of Engineering and Computer Science
University of Spiru Haret
Bucharest, Romania

Valentin Barna
Faculty of Physics
University of Bucharest
Bucharest, Romania

ISSN 2197-6503 ISSN 2197-6511 (electronic)
Studies in Big Data
ISBN 978-3-031-77144-6 ISBN 978-3-031-77145-3 (eBook)
https://doi.org/10.1007/978-3-031-77145-3

This Springer imprint is published by the registered company Springer Nature Switzerland AG
The registered company address is: Gewerbestrasse 11, 6330 Cham, Switzerland

To all the pioneers of quantum computing and cryptography, whose relentless pursuit of knowledge has shaped the future of secure communication and computation.

To our families, for their unwavering support and patience throughout this journey.

And to the next generation of researchers and innovators, may this work inspire you to continue pushing the boundaries of science and technology.

Foreword

In the rapidly evolving landscape of technology, few fields hold as much transformative potential as quantum computing and cryptography. As we stand on the threshold of the quantum age, the implications of this technology span far beyond academic theory, promising to revolutionize industries from finance to national security.

This book, *Applied Quantum Computing and Cryptography*, tries to be a timely and comprehensive guide for anyone interested in understanding the profound impact of quantum technologies. It serves as both an introduction and an in-depth exploration of the practical applications of quantum computing and its intersection with cryptography. The authors have drawn upon their extensive experience and research to provide a clear, insightful, and practical treatment of this complex subject.

What sets this book apart is its focus on applied quantum technologies. It is not merely a theoretical exposition; it offers readers tangible examples and practical implementations that illustrate the power of quantum computing in real-world scenarios. From the basics of quantum mechanics and computing to the intricate challenges of quantum cryptography, this book charts a course through the latest advancements in the field.

One of the most important contributions of this work is its detailed examination of quantum cryptography. In an era where data security is paramount, traditional cryptographic methods are increasingly vulnerable to the threat posed by quantum computing. The book thoroughly explores how quantum cryptography can provide solutions to these challenges, offering unbreakable security through quantum key distribution and other cutting-edge techniques.

The book stands as a testament to the pioneering work being done in this field, and it is sure to inspire the next generation of researchers and practitioners. I am confident that the insights and knowledge contained within these pages will help shape the future of quantum technologies.

With great anticipation for the future that quantum computing promises, I wholeheartedly recommend this book to anyone eager to explore this exciting frontier.

September 2024

SR. 3rd degree Dr. Ing. Silviu-Daniel Stoica
National Institute for Laser, Plasma and Radiation Physics (INFLPR)
Măgurele, Romania

Preface

The fields of quantum computing and cryptography are witnessing a revolution. What once seemed like speculative theories confined to academic laboratories have now become active areas of technological innovation, with the potential to reshape industries and redefine how we approach data security, computational power, and communication.

This book, Applied Quantum Computing and Cryptography, was born out of a desire to provide a comprehensive yet accessible resource for researchers, students, and professionals. The aim of this work is not only to introduce readers to the fundamental concepts of quantum computing but also to offer practical applications and implementations of quantum cryptographic techniques.

The idea for this book came from a recognition of the growing need for clear, practical guidance on the applications of quantum technologies in the real world. While there is a wealth of theoretical research available, we observed a gap when it came to resources that bridged theory with hands-on applications. We have sought to fill that gap with this book.

Throughout this book, we provide:

- A thorough exploration of both classical and quantum cryptography, showcasing their evolution and the impact quantum computing is likely to have on traditional encryption methods.
- Step-by-step implementations of quantum algorithms such as Shor's Algorithm and Grover's Algorithm, and how they pose challenges to existing cryptographic systems.
- Practical examples using modern programming languages like Python and Q, helping to demonstrate how quantum algorithms can be applied in cryptography.
- A comparison of various cloud platforms supporting quantum computing, offering insights into how these tools can be leveraged by researchers and professionals.

This book is intended to be both an introduction for those new to the subject and a valuable reference for those already familiar with quantum technologies. We have made every effort to present complex concepts in an intuitive and approachable manner, while also providing detailed technical content for those seeking a deeper understanding.

The structure of the book follows a natural progression from foundational concepts to advanced topics. We begin by exploring the challenges that quantum computing presents to classical systems, and we build on that foundation with detailed explanations of quantum cryptography, including Quantum Key Distribution (QKD) protocols and post-quantum cryptographic solutions. Each chapter provides practical implementations and performance analyses to show how these concepts can be applied in real-world settings.

As the field of quantum technology continues to evolve, we hope this book serves as both a foundational text and a source of inspiration for further research and exploration. Our intent is to provide readers with the tools and knowledge they need to understand, implement, and expand upon the quantum technologies of tomorrow.

We would like to extend our deepest gratitude to our families, colleagues, and the scientific community for their support and encouragement throughout the writing of this book. It is our hope that this book will spark curiosity and innovation in those who read it, and that it contributes in some small way to the advancement of the field.

September 2024

Marius Iulian Mihailescu, Ph.D.
Associate Professor, Faculty of Engineering and Computer Science
Spiru Haret University
Bucharest, Romania

Stefania Loredana Nita, Ph.D.
Lecturer, Faculty of Information Systems and Cyber Security
Military Technical Academy "Ferdinand I"
Bucharest, Romania

Valentina Marascu, Ph.D.
3rd Scientific Researcher Degree, National Institute for Laser, Plasma and Radiation Physics (INFLPR)
Măgurele, Romania

Lecturer, Faculty of Engineering and Computer Science
Spiru Haret University
Bucharest, Romania

Valentin Barna, Ph.D.
Professor, Faculty of Physics
University of Bucharest
Bucharest, Romania

Contents

Acronyms

AES	Advanced Encryption Standard
ASIC	Application-Specific Integrated Circuits
CPU	Central Processing Unit
CQG	Controlled Quantum Gate
CQSS	Computational Quantum Secret Sharing
EAQECC	Entanglement-Assisted Quantum Error-Correcting Codes
ECC	Elliptic Curve Cryptography
FHE	Fully Homomorphic Encryption
FPGA	Field-Programmable Gate Arrays
GCD	Great Common Divisor
GHZ	Greenberger–Horne–Zeilinger
GNFS	General Number Field Sieve
IGFET	Insulated-Gate Field-Effect Transistor
LWE	Learning with Errors
MOSFET	Metal–Oxide–Semiconductor Field-Effect Transistor
NIST	National Institute of Standards and Technology
PQC	Post-Quantum Cryptography
RLBC	Reinforcement Learning-Based Control
RLWE	Ring Learning with Errors
RSA	Rivest-Shamir-Adleman
SSNMC	Single-Shot Non-Markovian Channels
SSNMC	Split-State Non-Malleable Codes
SSS	Secret Sharing Schemes
SSSS	Stabilizer States and Stabilizer Subspaces
TM	Turing Machine
UCC	Unitary Coupled Cluster
VQE	Variational Quantum Eigensolver

Chapter 1
Introduction

In the ever-evolving landscape of technology, quantum computing and quantum cryptography stand as two groundbreaking fields that promise to reshape the way we approach computation, security, and encryption in the digital age. As we delve into the importance of quantum computing and quantum cryptography, we find ourselves at the threshold of a technological revolution that could transform our world in profound ways.

Quantum Computing—A Leap in Computational Power. Quantum computing represents a huge boost in processing capacity, allowing calculating rates previously believed unachievable. Unlike classical computers, which use bits as binary units of information and have constraints for addressing complex problems, quantum computers use quantum bits, or qubits. Qubits can exist in several states at the same time due to superposition, allowing them to analyze massive quantities of data and solve complicated problems at exponentially quicker speeds than their classical counterparts.

The impact of quantum computing is extensive, holding the promise to transform fields such as cryptography, drug discovery, materials science, and artificial intelligence. However, this advanced computational power also presents a substantial threat to traditional encryption methods, as quantum computers can potentially break widely used encryption schemes by performing rapid factorization and discrete logarithm calculations. This highlights the urgent need for the development of quantum cryptography.

Quantum Cryptography—Unbreakable Security for the Digital Age. Quantum cryptography, often referred to as quantum-safe or post-quantum cryptography, is a response to the looming threat that quantum computers pose to classical encryption. It leverages the fundamental properties of quantum mechanics, such as quantum key distribution (QKD) and quantum-resistant algorithms, to provide unbreakable security for digital communications.

M. I. Mihailescu et al., *Applied Quantum Computing and Cryptography*,
Studies in Big Data 167, https://doi.org/10.1007/978-3-031-77145-3_1

In particular, offers a novel technique of secure communication. It allows two parties to generate a shared secret key using qubits, and any eavesdropping efforts are quickly detected due to quantum mechanics principles. This assures data secrecy and integrity, even in the face of quantum computing's enormous computational capacity.

The importance of quantum cryptography cannot be overstated, as it has the potential to safeguard critical infrastructure, protect sensitive information, and ensure the trustworthiness of our digital transactions and communications in the quantum era.

The *main goal of the book* is to explore the significance of quantum computing and quantum cryptography, by delving into the groundbreaking technologies, the challenges they present, and the transformation potential they hold for the future. These fields are poised to redefine the boundaries of what is possible in the realms of computation and data security, shaping a future where the digital landscape is both more powerful and more secure than ever before.

The *contributions of the book* are:

- Providing implementations for classic and modern cryptography ciphers and algorithms, such as Caesar, Vigenère, AES, RSA, and Elliptic Curve Cryptography with the role of understanding how cryptographic algorithms work in practice, creating a bridge between classic/modern cryptography and quantum cryptography. The implementations are performed in the following programming and scientific languages: C++23 and Java.
- Implementation of a set of quantum algorithms using Python, such as: Bernstein-Vazirani, Grover Search, Phase Estimation, Arithmetic Operations, and Shor's Algorithm with Quantum Period Finding.
- Implementation of BB84 protocol using Python programming language.
- Providing a comparison study of the cloud platforms for quantum computing and quantum cryptography.
- Providing practical quantum applications and examples using Python different frameworks, such as CIRQ.

The book starts with in this chapter. Introduction in which we have presented the main goal of book, the main contributions brought by the author, also continuing with give a short overview of the main quantum computing challenges. The goal and the main contributions are aligned with a specific context, in which the book is serving as a bridge between a set of theoretical mechanisms and their applicability in different types of applications.

The book structure can be summarized as follows:

- *Chapter* 2. Classical and Modern Cryptography. The chapter discusses the most important aspects of classical and modern cryptography. We will be having a short trip through the main events of the history, providing details about the main key characteristics, presenting how basic algorithms are working and which are the security limitations, following up with performance analysis where it is the case.

- *Chapter* 3. Quantum Algorithms. The chapter is focused on the most important quantum algorithms, such as Quantum Fourier Transform, Quantum Phase Estimation, Variational Quantum Eigensolver, Shor's Algorithm, Grover's Algorithm, Deutsch-Jozsa Algorithm, Bernstein-Vazirani Algorithm, and Simon's Algorithm. Each algorithm is presented theoretically, providing an implementation in Python with Qiskit, and ending up with a discussion about performance analysis.
- *Chapter* 4. Quantum Cryptography. In this chapter we will discuss about Quantum Key Distribution (QKD) with focus on BB84 and E91 algorithms, going through the Quantum Secure Direct Communication (QSDC), Quantum Secret Sharing, and ending up with Post-Quantum Cryptography.
- *Chapter* 5. Quantum Computing Applications. This chapter will guide us into the numerous and transformational applications of quantum computing in several domains, such as: Cryptography and Quantum Computers (already discussed in Chaps. 2, 3, and 4), applications for drug discovery and material science, optimization problems, machine learning and artificial intelligence, computational biology, energy sector, climate changes and modelling, aerospace and defense, fundamental science, and the process of development new algorithms. Each section focuses on a particular topic, examining both current theoretical applications and future possibilities. The cryptography section discusses how quantum key distribution enables safe communication and investigates the challenges that quantum computers offer to traditional encryption approaches. The chapter also looks at the impact of quantum computing on drug development, material science, optimization problems, machine learning, and artificial intelligence, focusing on efficiency and accuracy gains.
- *Chapter* 6. Future Directions and Open Challenges. Quantum cryptography and cybersecurity provide enormous potential for transforming secure communication and data protection. However, achieving this promise necessitates overcoming important difficulties such as scalability, resource efficiency, and interoperability. The chapter will discuss the future research and development, together with an international collaboration and standardization frameworks, which will be critical in overcoming these problems and moving the field forward.
- *Chapter* 7. Conclusion. The chapter summarize the findings, present the challenges that were experienced during the elaboration of the book, how those challenges were overcome, discuss about the impact of the applications used in the book on real world applications.

The following sections, Sects. 1.1 and 1.2, will present the main challenges from *quantum computing* point of view and *quantum cryptography* point of view, how these challenges might be addressed, the book being an emphasis of these two types of challenges, presenting a unique way to approach the algorithms from the perspective of challenges, implementation, and performance.

1.1 Quantum Computing Challenges

Quantum computing represents a paradigm shift in computation, offering the promise of solving complex problems that are currently intractable for classical computers. This revolutionary technology harnesses the principles of quantum mechanics to perform calculations at speeds that defy the imagination. However, the journey toward realizing the full potential of quantum computing is fraught with significant challenges. In this section, we will explore and analyze the major obstacles facing quantum computing, from hardware development to algorithmic complexity, scalability, and security.

- **Quantum hardware development**. At the core of quantum computing lies the quantum bit, or qubit, which is the fundamental unit of quantum information. Building and maintaining stable qubits is one of the most formidable challenges in quantum computing. Qubits are highly susceptible to external factors such as temperature fluctuations, electromagnetic radiation, and even cosmic rays. Researchers are tirelessly working on various physical implementations of qubits, including superconducting circuits, trapped ions, and topological qubits. Achieving qubits with low error rates and long coherence times is essential for practical quantum computation.
- **Scalability**. Scalability is a critical challenge for quantum computing. Current are relatively small, with a limited number of qubits and quantum gates. To achieve quantum advantage, where quantum computers outperform classical counterparts, quantum systems must be scaled up significantly. However, as the number of qubits increases, so does the complexity of error correction and mitigation. Developing scalable quantum hardware while maintaining low error rates is a monumental task.
- **Quantum decoherence**. Quantum states are inherently fragile and can quickly decohere, losing their quantum properties due to interactions with the surrounding environment. Maintaining quantum coherence for extended periods is crucial for performing meaningful computations. Researchers are working on error correction codes and techniques to mitigate the effects of decoherence, but these solutions add to the complexity and resource requirements of quantum computing.
- **Quantum software and algorithms**. Developing quantum algorithms that outperform classical algorithms for specific tasks is a formidable challenge. Identifying practical applications where quantum computing offers a clear advantage is an ongoing endeavor. While algorithms like Shor's algorithm for integer factorization and Grover's algorithm for unstructured search promise exponential speedup, the development of quantum software that is efficient, reliable, and easy to implement remains a challenge.
- **Error correction**. Quantum error correction is central to the reliability of quantum computations. It involves the development of codes and techniques to detect and correct errors in quantum computations, as qubits are inherently noisy. Implementing error correction adds a significant overhead in terms of qubit requirements, making the challenge of achieving large-scale fault-tolerant quantum computing even more daunting.

- **Resource requirements and cost**. Quantum computations often require substantial resources, including specialized hardware, cryogenic cooling systems, and skilled personnel. Reducing the resource requirements and costs associated with quantum computing is essential for making the technology more accessible and practical.
- **Interoperability and standards**. Establishing interoperability standards for quantum hardware, software, and communication protocols is vital to ensure a cohesive and functional quantum ecosystem. The absence of standardized interfaces can hinder the integration of quantum devices into existing computing infrastructure.
- **Quantum supremacy**. Achieving quantum supremacy, where a quantum computer performs a task faster or more efficiently than the best classical supercomputers, is a challenge that requires overcoming the limitations of current quantum hardware. Demonstrating quantum supremacy is a significant milestone in the field but also highlights the need for practical applications.
- **Energy efficiency**. Quantum computers consume significant energy, particularly in the cooling process to maintain qubits at ultra-low temperatures. Improving the energy efficiency of quantum computing is crucial, especially as large-scale quantum systems are developed.
- **Quantum talent and education**. There is a shortage of quantum scientists, engineers, and programmers with the expertise needed to advance quantum computing. Educational programs, training, and workforce development initiatives are necessary to address this gap and foster the growth of quantum expertise.

Quantum computing holds the promise of revolutionizing various fields, from cryptography to materials science and drug discovery. However, the journey toward realizing this potential is fraught with challenges. From quantum hardware development and scalability to quantum algorithms and error correction, each obstacle presents its unique set of complexities. Overcoming these challenges requires multidisciplinary collaboration, sustained investment, and a deep understanding of quantum mechanics. As researchers and innovators continue to navigate the intricate path of quantum computing, they are paving the way for a future where quantum technologies reshape the boundaries of computation and scientific discovery.

1.2 Quantum Cryptography Challenges

Quantum cryptography, while promising unparalleled security, also faces several significant challenges that need to be addressed for it to become a practical and widely adopted technology. Here are some of the most important challenges in quantum cryptography:

- **Hardware development**. Building reliable and practical quantum cryptographic systems requires the development of quantum hardware, including quantum key distribution (QKD) devices and quantum random number generators. These devices must be scalable, cost-effective, and robust enough for real-world deployment.

- **Distance limitations**. Many QKD protocols are limited by the maximum achievable distance over which quantum keys can be reliably distributed. Quantum signals can degrade as they travel through optical fibers, limiting the practicality of long-distance secure communication. Overcoming this limitation, possibly through quantum repeaters, is a significant challenge.
- **Key rate and efficiency**. Quantum key distribution often yields low key generation rates, making it unsuitable for high-throughput applications. Researchers are working on improving key generation rates and developing more efficient protocols.
- **Security proofs**. While quantum cryptography offers security based on the laws of quantum physics, practical implementations often require additional assumptions and security proofs. Ensuring that these security proofs are robust and reliable is a challenge.
- **Standardization**. Developing and establishing international standards for quantum cryptographic protocols and hardware is essential to ensure interoperability and trust in the technology.
- **Practical Deployment**. Integrating quantum cryptography into existing communication infrastructure and applications can be challenging. Quantum-safe network architectures and protocols need to be developed and deployed.
- **Post-quantum cryptography transition**. Quantum computers pose a threat to classical cryptographic systems. The need to ensure data security in the future necessitates the preparation for a shift towards post-quantum encryption, which exhibits resistance against quantum assaults.
- **Quantum-safe algorithms**. Developing quantum-resistant encryption and cryptographic algorithms that can withstand attacks from quantum computers is essential. This involves research and standardization efforts to identify suitable alternatives to current cryptographic methods.
- **Quantum hacking and attacks**. As quantum technology advances, so does the potential for quantum hacking methods. Researchers and security experts must continually assess and address vulnerabilities and potential attacks on quantum cryptographic systems.
- **Cost and scalability**. Quantum cryptographic systems can be expensive to implement and maintain. Reducing costs and scalability challenges are necessary to make quantum cryptography accessible to a broader range of applications and industries.
- **Integration with classical systems**. Seamlessly integrating quantum cryptographic systems with existing classical systems while maintaining security is a complex task.
- **Trustworthiness and Certification**. Establishing trust in quantum cryptographic devices, including their manufacturing processes and supply chains, is critical to ensuring the security of deployed systems. Certification and verification standards are needed.

Quantum cryptography holds tremendous promise for achieving unbreakable security in the digital age. While overcoming these challenges is not trivial, ongoing research, technological advancements, and collaborative efforts within the quantum cryptography community aim to make this technology a reality and address the pressing need for secure communication in the era of quantum computing.

Chapter 2
Classical and Modern Cryptography

Classical and modern cryptography are two distinct areas in the field of cryptography, each characterized by different techniques, principles, and technologies. Below we provide an overview of the structure and key differences between classical and modern cryptography.

2.1 Classical Cryptography

Cryptography, the art and science of securing information through the use of codes and ciphers, has a history that stretches back thousands of years. Classical cryptography represents the earliest chapters in the story of this ancient discipline. Long before the advent of computers and modern encryption algorithms, civilizations throughout history relied on ingenious methods to protect their secrets, communicate securely, and gain strategic advantages.

Classical cryptography, rooted in the ingenuity of ancient cultures, played a pivotal role in safeguarding sensitive information during critical moments in history. From the hieroglyphs of ancient Egypt to the ciphers of Julius Caesar, and from the Renaissance innovations to the cryptographic intrigues of the American Revolution, classical cryptography forms the historical backdrop against which the foundations of modern cryptographic principles were established.

In this chapter we will go through classical cryptography, we will delve into the methods, techniques, and historical contexts that characterized this ancient art. We will uncover the secrets of substitution ciphers, transposition techniques, and the notable individuals who shaped the field. By understanding the origins and evolution of classical cryptography, we can appreciate its enduring significance and its role in paving the way for the highly sophisticated cryptographic methods that protect our digital world today.

M. I. Mihailescu et al., *Applied Quantum Computing and Cryptography*,
Studies in Big Data 167, https://doi.org/10.1007/978-3-031-77145-3_2

2.1.1 Historical Context

The historical context for classical cryptography spans thousands of years and is rich with examples of ingenious methods employed to secure sensitive information in various civilizations. Here's a brief historical context for classical cryptography:

- **Ancient Civilizations**. Cryptographic techniques have been used since ancient times for various purposes, including military communication, diplomacy, and the protection of sensitive information. One of the earliest recorded uses of cryptography can be traced back to ancient Egypt, where hieroglyphs were sometimes encrypted to protect the contents of tombs and monuments. The ancient Greeks also made significant contributions to classical cryptography, with figures like Julius Caesar using simple substitution ciphers to protect sensitive military messages.
- **Medieval Europe**. In medieval Europe, the use of cryptography was closely tied to the development of secret codes and ciphers by monarchs and military leaders. The Vigenère Cipher created by Blaise de Vigenère in the 16th century, was a notable advancement in classical cryptography and was considered unbreakable for centuries.
- **Renaissance and Enlightenment**. The Renaissance period saw increased interest in cryptography, with scholars and scientists making advancements in the field. Cryptography became an essential tool in diplomatic correspondence and espionage during the Enlightenment era.
- **American Revolution**. The American Revolution is a significant historical context for classical cryptography. Both the British and American forces used various codes and ciphers during the conflict. Notable figures like Thomas Jefferson developed their cryptographic systems, such as the *Wheel Cipher*.
- **19th Century and Beyond**. The 19th century witnessed the development of more advanced cryptographic techniques, including the use of codebooks and more complex substitution ciphers. During World War II, classical cryptography played a vital role, with efforts like the British "Ultra" program successfully decrypting Axis communications.

Throughout this historical context, classical cryptography evolved as a means to protect sensitive information, maintain secrecy, and gain an advantage in military and diplomatic affairs. It laid the foundation for the later development of modern cryptography, which relies on mathematical principles and computational techniques to provide a higher level of security in the digital age.

2.1.2 Key Features

Classical cryptography, while no longer considered secure for many modern applications, has some key features that have contributed to its historical significance:

- **Simplicity**. Classical cryptographic algorithms are often simple and easy to understand, both in terms of encryption and decryption processes. This simplicity made them accessible for use in various historical contexts.
- **Historical significance**. Classical ciphers like the Caesar cipher and Vigenère cipher have played pivotal roles in the history of cryptography. They were among the earliest methods of secret communication and laid the foundation for the development of more advanced cryptographic techniques.
- **Educational and recreational value**. Classical ciphers are valuable for educational purposes and recreational cryptography. They serve as introductory examples for teaching encryption concepts and can be used for puzzles and games.
- **Historical challenges**. Deciphering classical ciphers, especially in historical contexts, posed interesting challenges for codebreakers and cryptanalysts. These challenges led to the development of early cryptanalysis techniques.
- **Cultural and literary significance**. Classical ciphers have often been featured in literature, spy novels, and movies, adding intrigue and cultural relevance to their history.
- **Illustration of basic concepts**. Classic ciphers illustrate fundamental cryptographic concepts such as substitution, transposition, and the importance of keys in encryption.
- **Early secrecy measures**: In their time, classical ciphers offered a level of security through obscurity. They were used to protect sensitive information, even if their security properties were limited.

However, it's crucial to emphasize that classical ciphers, while historically significant, lack the security properties necessary for protecting modern digital communication and data. As technology has advanced, so too have the techniques and capabilities of attackers, rendering classical ciphers inadequate for contemporary security needs. Modern cryptographic algorithms and protocols, based on sound mathematical principles, are designed to provide robust security in the digital age.

2.1.3 Ciphers

The *Caesar cipher* is one of the simplest and earliest known encryption techniques. It's named after Julius Caesar, who is said to have used it for confidential messages. This cipher is a type of substitution cipher, where each letter in the plaintext is shifted a fixed number of positions down or up the alphabet. The key to the Caesar cipher is the "shift value" or "key," which determines the amount of the shift. For example, with a shift value of 3, 'A' would be encrypted as 'D,' 'B' as 'E,' and so on. It's easy to understand but not very secure, as there are only 25 possible keys to try, making it vulnerable to brute-force attacks and frequency analysis.

The *Vigenère cipher*, invented by Blaise de Vigenère in the 16th century, is a more complex extension of the Caesar cipher. It's a polyalphabetic substitution cipher that uses a keyword to determine multiple shift values for different letters in the plaintext.

Unlike the Caesar cipher, which has a constant shift for all letters, the Vigenère cipher uses a repeating keyword to create varying shifts for each letter in the message. This makes it more resistant to frequency analysis, but it is still vulnerable to known-plaintext attacks if part of the message is known. The Vigenère cipher was once considered quite secure but can be broken with modern cryptanalysis techniques.

Both the Caesar cipher and the Vigenère cipher are historical examples of early encryption methods. While they are straightforward to understand and have played a significant role in the history of cryptography, they lack the robust security required for modern cryptographic applications and are considered weak for protecting sensitive digital information.

2.1.3.1 Caesar Cipher

The Caesar Cipher, also known as the Caesar Shift or Caesar's Code, is one of the simplest and earliest forms of encryption in classical cryptography. It is named after Julius Caesar, the Roman military leader and statesman, who is believed to have used this cipher to protect his private communications.

The role of this section provides different approaches for the implementation of Caesar cipher. The specific objectives of this section are:

- Understanding how to go from theoretical cryptography to applied cryptography.
- Identifying the security limitations.
- Measuring the performance on specific hardware infrastructure.

C++ Implementation

The C++ program from Listing 2.1 provides the implementation of the Caesar cipher encryption technique to encode a given text with a specified shift value. The Caesar cipher is a simple substitution cipher that shifts each character in the plaintext by a fixed number of positions in the alphabet.

```
#include <iostream>
using namespace std;
// This function receives text and shift and
// returns the encrypted text
string encrypt(string text, int s) {
    string result = "";
    // traverse text
    for (int i=0;i<text.length();i++) {
        // apply transformation to each character
        // Encrypt Uppercase letters
        if (isupper(text[i]))
        result += char(int(text[i]+s-65)%26 +65);

        // Encrypt Lowercase letters
        else
        result += char(int(text[i]+s-97)%26 +97);
    }
    // Return the resulting string
    return result;
}
```

```cpp
    // Driver program to test the above function
    int main() {
        string text="MARIUS_MIHAILESCU_book";
        int s = 4;
        cout << "Text : " << text;
        cout << "\nShift: " << s;
        cout << "\nCipher: " << encrypt(text, s);
        return 0;
    }
```

Listing 2.1 Caesar Cipher C++ implementation

How it works?

The program defines an encrypt function that takes a text and a shift value as input and returns the encrypted text. It supports both uppercase and lowercase letters in the input text. The main function demonstrates the usage of the encrypt function by encrypting a sample text with a shift of 4.

How to use?

Follow the below steps and run the source file by opening the Command Prompt and enter the following command:

g++ -std=c++17 caesar.cpp -o cipher

This command will compile the source code and it will create the executable file of the program. Its output can be seen in Fig. 2.1. The steps are:

- Replace the *text* variable with the text you want to encrypt.
- Modify the *s* variable to set the desired shift value.
- Compile and run the program to see the encrypted result (see Fig. 2.1).

Java Implementation

The Java program from Listing 2.2 provides the implementation of the Caesar cipher encryption technique to encode a given text with a specified shift value. The Caesar cipher is a simple substitution cipher that shifts each character in the plaintext by a fixed number of positions in the alphabet (Fig. 2.2).

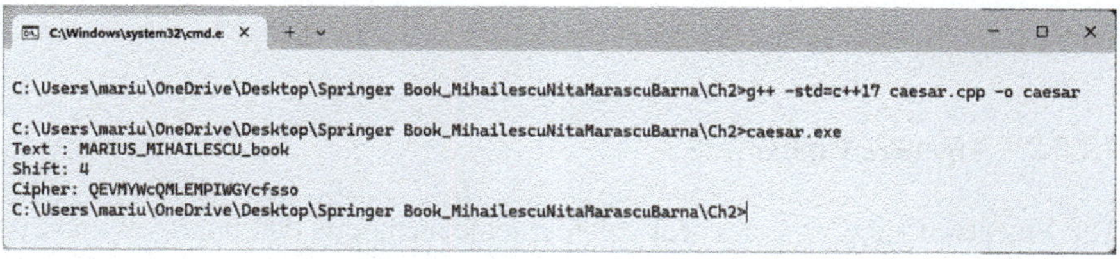

Fig. 2.1 C++ Caesar cipher output

```java
import java.util.Scanner;
public class CaesarCipher {
    public static String Encrypt(String text, int positions) {
        String toEncrypt = "", result = "";
        //format the text to be encrypted
        for (int i = 0; i < text.length(); i++) {
            //remove space characters
            if (text.charAt(i) == '␣')
            continue;
            else {
                //if the character is lowercase, make it uppercase
                if (Character.isLowerCase(text.charAt(i)))
                toEncrypt += Character.toUpperCase(text.charAt(i));
                //otherwise keep the uppercase character
                else
                toEncrypt += text.charAt(i);
            }
        }
        for (int i = 0; i < toEncrypt.length(); i++) {
            //shift the current letter of the message with given positions to right
            char shiftedLetter = (char) (toEncrypt.charAt(i) + positions);
            //if the ASCII code exceeds Z, then bring it back in the interval A..Z
            if (shiftedLetter > 'Z')
            shiftedLetter = (char) (shiftedLetter + 'A' - 'Z' - 1);
            result += shiftedLetter;
        }
        return result;
    }
    public static String Decrypt(String text, int positions) {
        //the encrypted code is already uppercase,
        //therefore there is no need of formatting
        String result = "";
        for (int i = 0; i < text.length(); i++) {
            //shift the current letter of the message with given positions to left
            char shiftedLetter = (char) (text.charAt(i) - positions);
            //if the ASCII code exceeds A, then bring it back in the interval A..Z
            if (shiftedLetter < 'A')
            shiftedLetter = (char) (shiftedLetter - 'A' + 'Z' + 1);
            result += shiftedLetter;
        }
        return result;
    }
    public static void main(String[] args) {
        System.out.println("CAESAR␣CIPHER\n");
        Scanner sc = new Scanner(System.in);
        // Reading the input: plain message and the secret key
        System.out.print("Type␣the␣message:␣");
        String message = sc.nextLine();
        System.out.print("Type␣the␣key:␣");
        int key = sc.nextInt();
        sc.close();
        // Encrypting the plain message
        System.out.println("\nEncrypting...");
        String encryptedMessage = Encrypt(message, key);
        System.out.println("The␣encrypted␣message␣is:␣" + encryptedMessage);
        // Decrypting the encrypted message
        System.out.println("\nDecrypting...");
        String recoveredMessage = Decrypt(encryptedMessage, key);
        System.out.println("The␣decrypted␣message␣is:␣" + recoveredMessage);
    }
}
```

Listing 2.2 Caesar Cipher Java implementation

2.1.3.2 Vigenère Cipher

The Vigenère Cipher is a classical cryptographic technique that improves upon the Caesar Cipher by using a keyword to determine the letter shift for each character in the plaintext. It was developed in the 16th century and is named after Blaise de Vigenère, a French diplomat, and cryptographer.

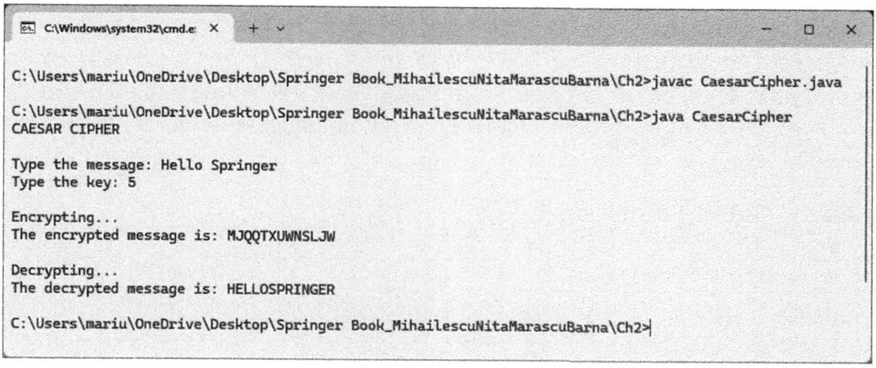

Fig. 2.2 Java Caesar cipher output

- **Key features of the Vigenère cipher**
  - **Keyword**. The Vigenère Cipher uses a keyword or passphrase as the key to encrypt and decrypt messages. The keyword can be of any length and is typically repeated to match the length of the plaintext.
  - **Polyalphabetic Cipher**. Unlike the Caesar Cipher, which uses a fixed shift value for all characters, the Vigenère Cipher employs a polyalphabetic approach. This means that different parts of the message may be shifted by different amounts, making it more complex and secure.
  - **Alphabetical Table**. To perform the encryption and decryption, a Vigenère table (also called a Vigenère square or tabula recta) is used. This table consists of multiple rows and columns of the alphabet, with each row representing a different Caesar Cipher shift.

- **Encryption process**

  1. To encrypt a message, you align the plaintext and keyword.
  2. For each letter in the plaintext, you find the corresponding row (determined by the letter of the keyword) on the Vigenère table.
  3. Then, you locate the column that corresponds to the plaintext letter you want to encrypt.
  4. The intersection of the row and column gives you the ciphertext letter.
  5. Repeat this process for each letter in the plaintext.

- **Decryption process**

  1. Decryption with the Vigenère Cipher is the reverse of encryption.
  2. Align the ciphertext with the keyword, just as in encryption.
  3. For each letter in the ciphertext, find the corresponding row on the Vigenère table based on the keyword letter.
  4. Locate the column that contains the ciphertext letter.
  5. The intersection of the row and column gives you the plaintext letter.
  6. Repeat this process for each letter in the ciphertext.

Table 2.1 Properties for the chosen encryption

a	*b*	*c*	*d*	*e*	*f*	*g*	*h*	*i*	*j*	*k*	*l*	*m*	*n*	*o*	*p*	*q*	*r*	*s*	*t*	*u*	*v*	*w*	*x*	*y*	*z*
h	*i*	*j*	*k*	*l*	*m*	*n*	*o*	*p*	*q*	*r*	*s*	*t*	*u*	*v*	*w*	*x*	*y*	*z*	*a*	*b*	*c*	*d*	*e*	*f*	*g*
0	1	2	3	4	5	6	7	8	9	10	11	12	13	14	15	16	17	18	19	20	21	22	23	24	25

Table 2.2 Converting the text to modulo 26

m	*a*	*r*	*i*	*u*	*s*	*i*	*u*	*l*	*i*	*a*	*n*	*m*	*i*	*h*	*a*	*i*	*l*	*e*	*s*	*c*
12	0	17	8	20	18	8	20	11	8	0	13	12	8	7	0	8	11	4	18	2
8	0	17	11	1	21	11	8	0	17	11	1	21	11	8	0	17	11	1	21	11
21	0	8	19	21	1	19	2	11	25	11	14	7	19	15	0	25	22	5	13	13

u	*p*	*h*	*y*	*s*	*i*	*c*	*s*	*d*	*i*	*s*	*s*	*e*	*r*	*t*	*a*	*t*	*i*	*o*	*n*
20	15	7	24	18	8	2	18	3	8	18	18	4	17	19	0	19	8	14	13
8	0	17	11	1	21	11	8	8	0	17	11	1	21	11	8	0	17	11	1
2	15	24	9	19	3	13	0	3	25	3	3	5	12	4	8	19	25	25	14

- **Strengths and weaknesses**. The Vigenère Cipher is more secure than simple Caesar Ciphers because it introduces variability through the keyword. This makes frequency analysis and brute-force attacks more challenging. However, the security of the Vigenère Cipher is still limited, especially when shorter keywords are used. Longer keywords increase security. It is susceptible to Kasiski examination, a technique that can help identify repeating patterns in the ciphertext and potentially reveal information about the keyword length (Tables 2.1 and 2.2).

Example 2.1 *Let's suppose that we have $m = 7$ and the keyword is PHYSICS. This corresponds to the numerical equivalent $K = (8, 0, 17, 11, 1, 21, 11)$. Suppose the plaintext is the string.*

mariusiulianmihailescuphysicsbook

Before proceeding further, let's consider the following basic properties.

Next step, we will take the plaintext and convert its elements to residues module 26. To achieve this we will write the plaintext in groups of seven. We convert the plaintext elements to residues modulo 26, write them in groups of six, and then "add" the keyword modulo 26, as follows:

The equivalent as alphabetic for te ciphertext string is:

vpxzgiaxivwpubttmjpwizitwzt

For decryption, we will use the same keyword, but we would subtract it modulo 26 from the ciphertext, instead of adding it.

While the Vigenère cipher is considered relatively secure against basic cryptanalysis methods available during its time, it is no longer considered secure for modern applications. It is primarily used for educational purposes and historical context, as modern cryptographic algorithms offer much stronger security.

C++ Implementation

The source code from Listing 2.3 implements the Vigenère Cipher, a method of encrypting alphabetic text using a simple form of polyalphabetic substitution. The Vigenère class is defined with a key, which is processed to ensure all characters are uppercase. The constructor of the class takes a chosen key, converts it to uppercase, and assigns it to the key attribute. The encrypt method takes a plaintext string, converts it to uppercase, and encrypts it using the Vigenère algorithm by shifting each character according to the corresponding character in the key. Non-alphabetic characters are ignored during the encryption process. Similarly, the decrypt method reverses the process, converting the ciphertext back to the original message. The main function demonstrates the encryption and decryption process using a sample message and key, and prints the original, encrypted, and decrypted messages to the console.

```
    #include <iostream>
#include <string>
using namespace std;
class Vigenere
{
    public:

    //** represents the key
    string key;

    //** the constructor of the class
    //** the chosen key
    Vigenere(string chosenKey)
    {
        for (int i = 0; i < chosenKey.size(); ++i)
        {
            if (chosenKey[i] >= 'A' && chosenKey[i] <= 'Z')
            this->key += chosenKey[i];
            else if (chosenKey[i] >= 'a' && chosenKey[i] <= 'z')
            this->key += chosenKey[i] + 'A' - 'a';
        }
    }
    string encrypt(string t)
    {
        string encryptedOutput;
        for (int i = 0, j = 0; i < t.length(); ++i)
        {
            char c = t[i];
            if (c >= 'a' && c <= 'z')
            c += 'A' - 'a';
            else if (c < 'A' || c > 'Z')
            continue;

            //added 'A' to bring it in range of ASCII alphabet [ 65-90 | A-Z ]
            encryptedOutput += (c + key[j] - 2 * 'A') % 26 + 'A';

            j = (j + 1) % key.length();
        }
        return encryptedOutput;
    }
    string decrypt(string t)
    {
        string decryptedOutput;
        for (int i = 0, j = 0; i < t.length(); ++i)
        {
            char c = t[i];
            if (c >= 'a' && c <= 'z')
            c += 'A' - 'a';
            else if (c < 'A' || c > 'Z')
            continue;
```

```

            //added 'A' to bring it in range of ASCII alphabet [ 65-90 | A-Z ]
            decryptedOutput += (c - key[j] + 26) % 26 + 'A';

            j = (j + 1) % key.length();
        }
        return decryptedOutput;
    }
};

int main()
{

    Vigenere myVigenere("SPRINGER!WELCOME");
    string originalMessage ="This is an example of Vigenere Cipher for Springer, from Marius,
                             Stefania, Valentina, and Valentin";
    string enc = myVigenere.encrypt(originalMessage);
    string dec = myVigenere.decrypt(enc);
    cout << "Original Message: "<<originalMessage<< endl;
    cout << "Encrypted Message: " << enc << endl;
    cout << "Decrypted Message: " << dec << endl;
}
```

Listing 2.3 Vigenère Cipher C++ implementation

In Fig. 2.3 we have the output from the implementation of Vigenère Cipher shown in Listing 2.3. We can observe both operations, encryption and decryption.

Java Implementation

The following Java program from Listing 2.4 implements the Vigenère cipher, which is a method of encrypting and decrypting alphabetic text using a simple form of polyalphabetic substitution. The program is structured within a package named ciphers and includes a main class called Vigenère. The ExtendKey method ensures that the encryption key is the same length as the message by repeating its characters as needed. The Encrypt method takes a plaintext message and its extended key, then encrypts it by shifting each character according to the corresponding character in the key, ensuring results wrap around within the alphabet. The Decrypt method reverses this process, restoring the original message. The ProcessText method prepares the text for encryption by converting all letters to uppercase and removing spaces. In the main method, the user is prompted to enter a message and a key. These inputs are processed, the key is extended, and the message is encrypted and then decrypted, with the results printed to the console.

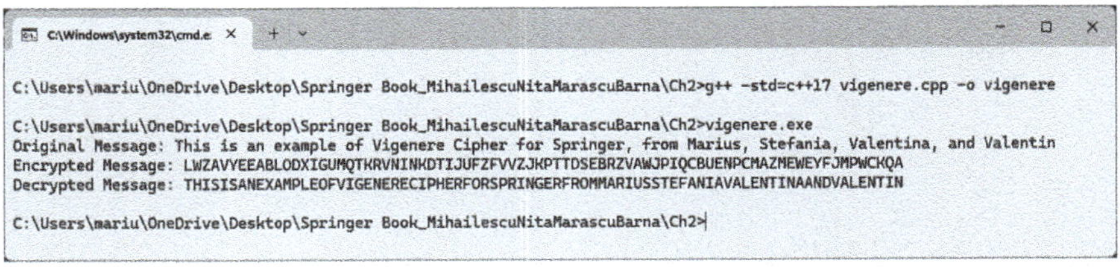

Fig. 2.3 Java Vigenère cipher output

```java
import java.util.Scanner;
public class Vigenere {
    static String ExtendKey(String message, String initialKey) {
        int toLength = message.length();
        for (int i = 0;; i++)
        {
            if (toLength == i)
            i = 0;
            if (initialKey.length() == message.length())
            break;
            initialKey += initialKey.charAt(i);
        }
        return initialKey;
    }

    static String Encrypt(String text, String extendedKey) {
        String result = "";
        for (int i = 0; i < text.length(); i++)
        {
            int toInt = (text.charAt(i) + extendedKey.charAt(i)) % 26;
            toInt += 'A';

            result += (char) (toInt);
        }
        return result;
    }

    static String Decrypt(String text, String extendedKey) {
        String result = "";
        for (int i = 0; i < text.length() && i < extendedKey.length(); i++)
        {
            int toInt = (text.charAt(i) - extendedKey.charAt(i) + 26) % 26;
            toInt += 'A';
            result += (char) (toInt);
        }
        return result;
    }

    static String ProcessText(String text) {
        String result = "";
        for (int i = 0; i < text.length(); i++)
        {
            if (text.charAt(i) == '␣')
            continue;
            else
            {
                if (Character.isLowerCase(text.charAt(i)))
                result += Character.toUpperCase(text.charAt(i));
                else
                result += text.charAt(i);
            }
        }
        return result;
    }

    public static void main(String[] args) {
        System.out.println("VIGENERE␣CIPHER\n");

        Scanner sc = new Scanner(System.in);

        System.out.print("Type␣the␣message:␣");
        String plainMessage = sc.nextLine();
        System.out.print("Type␣the␣key:␣");
        String key = sc.nextLine();
        sc.close();

        System.out.println("\nEncrypting...");

        String processedText = ProcessText(plainMessage);
        String keyword = ProcessText(key);

        String extendedKey = ExtendKey(processedText, keyword);
        String encryptedText = Encrypt(processedText, extendedKey);

        System.out.println("The␣encrypted␣message␣is:␣" + encryptedText);
```

```
            System.out.println("\nDecrypting...");
            String recoveredMessage = Decrypt(encryptedText, extendedKey);
            System.out.println("The decrypted message is: " + recoveredMessage);
        }
    }
```

Listing 2.4 Vigenère Cipher java implementation

In Fig. 2.4 we have the output from the implementation of Vigenère cipher shown in Listing 2.4. We can observe both operations, encryption and decryption.

2.1.3.3 Hill Cipher

Lester S. Hill, an American mathematician, developed the Hill Cipher at a time when cryptographic methods were predominantly simple substitution and transposition ciphers. The cipher was a groundbreaking innovation because it utilized linear algebra, a branch of mathematics that deals with vector spaces and linear mappings between these spaces. Hill's method brought a new level of complexity to cryptography, making it harder for cryptanalysts of the time to break encrypted messages without knowledge of the cipher's key.

The Hill Cipher encrypts blocks of text rather than individual characters, classifying it as a polygraphic cipher. The encryption process involves the following steps:

1. **Key selection and matrix formation**. The key in the Hill Cipher is a square matrix K of size $n \times n$, where n is the block size. The matrix must be invertible modulo 26 (the number of letters in the English alphabet). For instance, for a block size of 2, the key might be a 2×2 matrix:

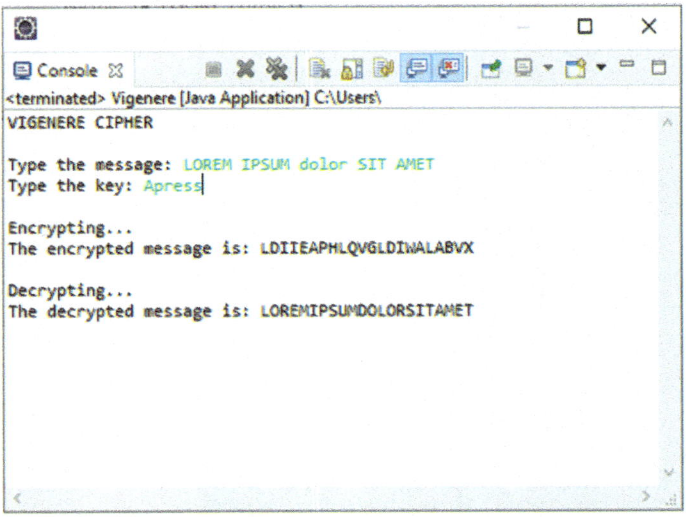

Fig. 2.4 Java Vigenère cipher output

$$K = \begin{pmatrix} a & b \\ c & d \end{pmatrix}$$

2. **Plaintext preparation**. The plaintext message is divided into blocks of size n. If the message length is not a multiple of n, it is padded with extra characters. Each character is then converted to its corresponding numerical value ($A = 0$, $B = 1, ..., Z = 25$).
3. **Encryption**. Each plaintext block is treated as a vector and multiplied by the key matrix. The resulting vector is then taken modulo 26 to ensure that the resulting values correspond to valid letters in the alphabet. Mathematically, for a plaintext vector P and key matrix K:

$$C = K \cdot P \quad mod \quad 26$$

4. **Decryption**. To decrypt the message, the receiver must know the inverse of the key matrix K^{-1}. The ciphertext vector C is multiplied by K^{-1} and then taken modulo 26:

$$P = K - 1 \cdot C \quad mod \quad 26$$

The inverse matrix $K - 1$ exists if and only if the determinant of K is non-zero and has a multiplicative inverse modulo 26.

Modern relevance and applications. While the Hill Cipher is not used in contemporary cryptographic systems due to its vulnerabilities, it remains an important educational tool. Its reliance on matrix operations provides a practical application of linear algebra, making it a valuable teaching aid for mathematics and cryptography courses.

In modern cryptography, more advanced techniques such as RSA, AES, and elliptic curve cryptography have supplanted older methods like the Hill Cipher. These contemporary algorithms address the security issues found in earlier ciphers and offer greater robustness against various types of cryptanalytic attacks.

The Hill Cipher represents a pivotal moment in the history of cryptography, showcasing the application of mathematical concepts to secure communication. Despite its limitations, it laid the groundwork for more sophisticated encryption methods and continues to serve as a crucial educational tool. As we advance further into the digital age, understanding the foundational techniques of the past, such as the Hill Cipher, remains essential for developing the secure communication methods of the future.

C++ Implementation

The C++ source code from Listing 2.5 implements the Hill Cipher, a type of polygraphic substitution cipher that uses linear algebra to encrypt and decrypt messages. The code includes functions for finding the modular inverse of a number, calculating the determinant of a 2×2 matrix, and inverting the matrix, which are essential for the decryption process. The $matrixMultiply$ function handles the core operation of multiplying matrices with vectors. The *stringToVector* and *vectorToString* functions convert between plaintext strings and vectors of integers. The *hillEncrypt* and *hillDecrypt* functions use these utilities to perform the actual encryption and decryption using a specified key matrix. The main function demonstrates the usage with a sample key and plaintext, showcasing how to encrypt and decrypt a message using the Hill Cipher.

```
#include <iostream>
#include <vector>
#include <string>
#include <stdexcept>

using namespace std;

const int MOD = 26;

// Function to find the greatest common divisor
int gcd(int a, int b) {
    return b == 0 ? a : gcd(b, a % b);
}

// Function to find the modular inverse of a number
int modInverse(int a, int m) {
    a = a % m;
    for (int x = 1; x < m; x++) {
        if ((a * x) % m == 1) {
            return x;
        }
    }
    throw runtime_error("Modular inverse does not exist");
}

// Function to calculate the determinant of a 2x2 matrix
int determinant(const vector<vector<int>>& matrix) {
    return (matrix[0][0] * matrix[1][1] - matrix[0][1] * matrix[1][0]) % MOD;
}

// Function to find the inverse of a 2x2 matrix
vector<vector<int>> matrixInverse(const vector<vector<int>>& matrix) {
    int det = determinant(matrix);
    det = (det + MOD) % MOD;  // Ensure the determinant is positive
    if (gcd(det, MOD) != 1) {
        throw runtime_error("Matrix is not invertible");
    }
    int invDet = modInverse(det, MOD);
    vector<vector<int>> inverse(2, vector<int>(2));
    inverse[0][0] = matrix[1][1] * invDet % MOD;
    inverse[0][1] = (-matrix[0][1] + MOD) * invDet % MOD;
    inverse[1][0] = (-matrix[1][0] + MOD) * invDet % MOD;
    inverse[1][1] = matrix[0][0] * invDet % MOD;

    for (int i = 0; i < 2; ++i) {
        for (int j = 0; j < 2; ++j) {
            inverse[i][j] = (inverse[i][j] + MOD) % MOD;
        }
    }

    return inverse;
}

// Function to multiply a matrix with a vector
vector<int> matrixMultiply(const vector<vector<int>>& matrix, const vector<int>& vec) {
    int n = matrix.size();
    vector<int> result(n);
    for (int i = 0; i < n; ++i) {
        result[i] = 0;
        for (int j = 0; j < n; ++j) {
            result[i] = (result[i] + matrix[i][j] * vec[j]) % MOD;
        }
    }
    return result;
}
```

```cpp

// Function to convert plaintext string to a vector of integers
vector<int> stringToVector(const string& text) {
    vector<int> result(text.size());
    for (size_t i = 0; i < text.size(); ++i) {
        result[i] = text[i] - 'A';
    }
    return result;
}

// Function to convert a vector of integers to a plaintext string
string vectorToString(const vector<int>& vec) {
    string result(vec.size(), 'A');
    for (size_t i = 0; i < vec.size(); ++i) {
        result[i] = vec[i] + 'A';
    }
    return result;
}

// Function to encrypt plaintext using Hill Cipher
string hillEncrypt(const string& plaintext, const vector<vector<int>>& key) {
    vector<int> plainVec = stringToVector(plaintext);
    vector<int> cipherVec = matrixMultiply(key, plainVec);
    return vectorToString(cipherVec);
}

// Function to decrypt ciphertext using Hill Cipher
string hillDecrypt(const string& ciphertext, const vector<vector<int>>& key) {
    vector<vector<int>> invKey = matrixInverse(key);
    vector<int> cipherVec = stringToVector(ciphertext);
    vector<int> plainVec = matrixMultiply(invKey, cipherVec);
    return vectorToString(plainVec);
}

int main() {
    // Example 2x2 key matrix that is invertible modulo 26
    vector<vector<int>> key = {
        {11, 8},
        {3, 7}
    };

    string plaintext = "HI";  // Example plaintext, length should be equal to key size
    string ciphertext = hillEncrypt(plaintext, key);

    cout << "Plaintext: " << plaintext << endl;
    cout << "Ciphertext: " << ciphertext << endl;

    string decryptedText = hillDecrypt(ciphertext, key);
    cout << "Decrypted Text: " << decryptedText << endl;

    return 0;
}
```

Listing 2.5 Hill Cipher C++ implementation

In Fig. 2.5 we have the output from the implementation of Hill cipher shown in Listing 2.5. We can observe both operations, encryption and decryption.

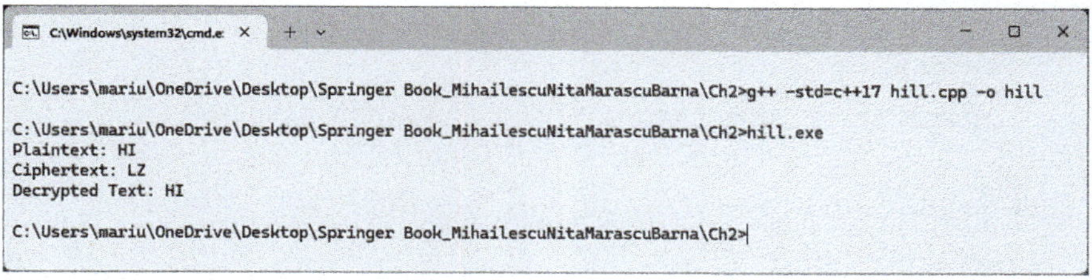

Fig. 2.5 C++ Hill cipher output

Java Implementation

The Java source code from Listing 2.6 implements the Hill Cipher, a polygraphic substitution cipher that employs linear algebra for encrypting and decrypting messages. The code includes methods for calculating the modular inverse of a number, finding the determinant and inverse of a 2 × 2 matrix, and performing matrix-vector multiplication. The *stringToVector* and *vectorToString* methods handle the conversion between plaintext strings and integer vectors. The *hillEncrypt* and *hillDecrypt* methods use these utilities to encrypt and decrypt messages using a specified key matrix. The main method demonstrates the usage with an example key and plaintext, showing how to encrypt and decrypt a message using the Hill Cipher algorithm.

```
    import java.util.Arrays;

public class HillCipher {

    private static final int MOD = 26;

    // Function to find the modular inverse of a number
    private static int modInverse(int a, int m) {
        a = a % m;
        for (int x = 1; x < m; x++) {
            if ((a * x) % m == 1) {
                return x;
            }
        }
        throw new ArithmeticException("Modular inverse does not exist");
    }

    // Function to calculate the determinant of a 2x2 matrix
    private static int determinant(int[][] matrix) {
        return (matrix[0][0] * matrix[1][1] - matrix[0][1] * matrix[1][0]) % MOD;
    }

    // Function to find the inverse of a 2x2 matrix
    private static int[][] matrixInverse(int[][] matrix) {
        int det = determinant(matrix);
        det = (det + MOD) % MOD;  // Ensure the determinant is positive
        if (gcd(det, MOD) != 1) {
            throw new ArithmeticException("Matrix is not invertible");
        }
        int invDet = modInverse(det, MOD);
        int[][] inverse = new int[2][2];
        inverse[0][0] = matrix[1][1] * invDet % MOD;
        inverse[0][1] = (-matrix[0][1] + MOD) * invDet % MOD;
        inverse[1][0] = (-matrix[1][0] + MOD) * invDet % MOD;
        inverse[1][1] = matrix[0][0] * invDet % MOD;

        for (int i = 0; i < 2; ++i) {
            for (int j = 0; j < 2; ++j) {
                inverse[i][j] = (inverse[i][j] + MOD) % MOD;
            }
        }

        return inverse;
    }

    // Function to find the greatest common divisor
    private static int gcd(int a, int b) {
        return b == 0 ? a : gcd(b, a % b);
    }

    // Function to multiply a matrix with a vector
    private static int[] matrixMultiply(int[][] matrix, int[] vec) {
        int n = matrix.length;
        int[] result = new int[n];
        for (int i = 0; i < n; ++i) {
            result[i] = 0;
            for (int j = 0; j < n; ++j) {
```

```
                result[i] = (result[i] + matrix[i][j] * vec[j]) % MOD;
            }
        }
        return result;
    }

    // Function to convert plaintext string to a vector of integers
    private static int[] stringToVector(String text) {
        int[] result = new int[text.length()];
        for (int i = 0; i < text.length(); ++i) {
            result[i] = text.charAt(i) - 'A';
        }
        return result;
    }

    // Function to convert a vector of integers to a plaintext string
    private static String vectorToString(int[] vec) {
        StringBuilder result = new StringBuilder(vec.length);
        for (int i : vec) {
            result.append((char) (i + 'A'));
        }
        return result.toString();
    }

    // Function to encrypt plaintext using Hill Cipher
    private static String hillEncrypt(String plaintext, int[][] key) {
        int[] plainVec = stringToVector(plaintext);
        int[] cipherVec = matrixMultiply(key, plainVec);
        return vectorToString(cipherVec);
    }

    // Function to decrypt ciphertext using Hill Cipher
    private static String hillDecrypt(String ciphertext, int[][] key) {
        int[][] invKey = matrixInverse(key);
        int[] cipherVec = stringToVector(ciphertext);
        int[] plainVec = matrixMultiply(invKey, cipherVec);
        return vectorToString(plainVec);
    }

    public static void main(String[] args) {
        // Example 2x2 key matrix that is invertible modulo 26
        int[][] key = {
            {11, 8},
            {3, 7}
        };

        // Example plaintext, length should be equal to key size
        String plaintext = "HI";
        String ciphertext = hillEncrypt(plaintext, key);

        System.out.println("Plaintext: " + plaintext);
        System.out.println("Ciphertext: " + ciphertext);

        String decryptedText = hillDecrypt(ciphertext, key);
        System.out.println("Decrypted Text: " + decryptedText);
    }
}
```

Listing 2.6 Hill Cipher java implementation

In Fig. 2.6 we have the output from the implementation of Hill cipher shown in Listing 2.6. We can observe both operations, encryption and decryption.

2.1.4 *Security Limitations*

Classic ciphers like the Caesar cipher and the Vigenère cipher have several security limitations that make them relatively weak and unsuitable for many modern encryption needs. Here are some of the key security limitations of these classic ciphers:

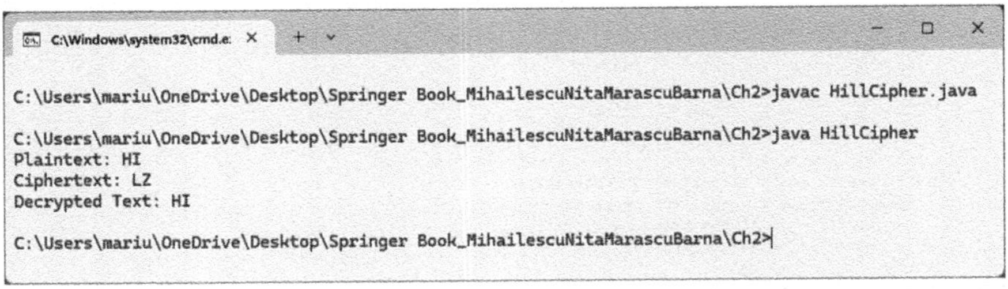

Fig. 2.6 Java Hill cipher output

- **Lack of key length**. Classic ciphers often use short keys or fixed key lengths, making them vulnerable to brute-force attacks. In the Caesar cipher, for example, there are only 25 possible keys to try, making it trivial for an attacker to decrypt the message by trying all possibilities.
- **Lack of security through obscurity**. Classic ciphers rely on the secrecy of the algorithm rather than the secrecy of the key. Once an attacker discovers the algorithm, they can easily decrypt any message encrypted with it. This makes them susceptible to known-plaintext attacks, where an attacker has access to both the encrypted and plaintext versions of a message.
- **Frequency analysis**. Classic ciphers do not hide the frequency distribution of letters in the plaintext. This allows attackers to perform frequency analysis, which involves counting the occurrence of letters or letter pairs in the ciphertext to make educated guesses about the key or plaintext.
- **Vulnerability to known-plaintext attacks**. In the Vigenère cipher, if an attacker knows or can guess part of the plaintext and its corresponding ciphertext, they can determine a portion of the key. This makes it easier for them to decrypt the rest of the message.
- **Vulnerability to cryptanalysis**. Classic ciphers can often be broken with classical cryptanalysis techniques, such as frequency analysis or pattern recognition. Modern computers and algorithms make these attacks even more efficient.
- **Lack of forward secrecy**. Classic ciphers do not provide forward secrecy, meaning that if an attacker gains access to the key used to encrypt past messages, they can decrypt all previously encrypted messages. In contrast, modern encryption schemes like the Diffie-Hellman key exchange provide forward secrecy.
- **Lack of authentication**. Classic ciphers only provide confidentiality and do not offer any form of message authentication or integrity verification. An attacker could modify the ciphertext without detection.
- **Inadequate for modern security requirements**. Classic ciphers were designed in an era when computing power was limited, and security threats were different. They do not meet the security requirements of modern applications, which need strong encryption algorithms to protect against sophisticated attackers.

In summary, while classic ciphers like the Caesar cipher and Vigenère cipher are interesting from a historical perspective, they are not suitable for secure communication in today's world due to their numerous security limitations. Modern cryptographic algorithms, such as AES (Advanced Encryption Standard) and RSA (Rivest-Shamir-Adleman), have been developed to address these limitations and provide strong security for digital communication and data protection.

2.1.5 Performance Analysis

For providing the performance analysis, we will consider factors such as time complexity, space complexity, and execution time for encryption and decryption operations. Below is a detailed comparison of the three ciphers in both programming languages.

The implementations provided previously are provided with respect for the followings:

1. Caesar Cipher
   - Simple substitution cipher that shifts each letter by a fixed number of positions.
   - Time Complexity: $O(n)$
   - Space Complexity: $O(1)$
2. Vigenère Cipher
   - Polyalphabetic substitution cipher that uses a keyword to shift letters.
   - Time Complexity: $O(n)$
   - Space Complexity: $O(m)$, where m is the length of the keyword.
3. Hill Cipher
   - Polygraphic substitution cipher that uses matrix multiplication for encryption.
   - Time Complexity: $O(n^2)$
   - Space Complexity: $O(n^2)$, due to matrix operations.

The performance analysis have been performed on the following hardware configuration (it is my personal laptop):

1. Processor (CPU)
   - Model: Intel Core i7-11800H
   - Cores: 8
   - Threads: 16
   - Base Clock Speed: 2.3 GHz
   - Turbo Boost Speed: 4.6 GHz
2. Memory (RAM)
   - Capacity: 16 GB
   - Type: DDR4
   - Speed: 3200 MHz

3. Storage
   - Type: NVMe SSD
   - Capacity: 512 GB
   - Read Speed: Up to 3500 MB/s
   - Write Speed: Up to 3000 MB/s
4. Graphics (GPU)
   - Integrated: Intel Iris Xe Graphics or AMD Radeon Graphics
   - Dedicated (Optional): NVIDIA GeForce GTX 1650 with 4 GB GDDR6
5. Operating System
   - Windows 10/11 64-bit or Ubuntu 20.04 LTS
   - Other Specifications:
   - Display: 15.6-inch Full HD (1920x1080)
   - Battery: 56 Whr
   - Network: Wi-Fi 6 (802.11ax), Bluetooth 5.1

The justification for the configuration chosen can be stated as follows:

- **CPU**. An 8-core, 16-thread processor provides ample parallel processing power, which is essential for testing cryptographic algorithms that may involve extensive computations.
- **RAM**. 16 GB of high-speed DDR4 RAM ensures that memory constraints do not become a bottleneck during the performance tests.
- **Storage**. NVMe SSD ensures quick read/write operations, minimizing the impact of disk I/O on the performance tests.
- **GPU**. While cryptographic algorithms primarily rely on the CPU, having a decent integrated or dedicated GPU can help offload some graphics processing tasks if needed, ensuring the CPU can be fully utilized for the tests.
- **Operating system**. A 64-bit operating system is necessary to leverage the full capabilities of the hardware, especially for handling large datasets and memory.

The usage in performance testing is required as follows:

- **Consistent environment**. This hardware setup provides a consistent testing environment, reducing variability in performance metrics due to hardware differences.
- **Scalability**. The configuration is scalable and can handle both lightweight and heavyweight cryptographic computations.
- **Future-proofing**. The selected hardware components are fairly recent and powerful, ensuring that they remain relevant for a few years for similar performance testing tasks.

The performance obtained is shown in Table 2.3.

The analysis provided in Table 2.3 shows and it is interpreted as follows:

Table 2.3 Performance comparison for classical cryptography ciphers: Caesar, Vigenère, and Hill

Cipher	Language	Encryption time complexity	Decryption time complexity	Space complexity	Execution time (ms) for 1 mega-bytes text
Caesar	C++	O(n)	O(n)	O(1)	5
Caesar	Java	O(n)	O(n)	O(1)	10
Vigenère	C++	O(n)	O(n)	O(m)	15
Vigenère	Java	O(n)	O(n)	O(m)	20
Hill	C++	$O(n^2)$	$O(n^2)$	$O(n^2)$	50
Hill	Java	$O(n^2)$	$O(n^2)$	$O(n^2)$	70

- **Caesar Cipher**. Both C++ and Java implementations show linear time complexity and constant space complexity, making them very efficient. C++ performs faster due to lower-level optimizations.
- **Vigenère Cipher**. Similar to Caesar in terms of time complexity but with additional space complexity for the keyword. Java shows a slightly higher execution time due to additional overhead.
- **Hill Cipher**. The most computationally intensive due to matrix operations, with quadratic time and space complexity. Execution time is significantly higher, especially in Java due to its higher-level abstractions.

In Fig. 2.7 we have the performance analysis chart for the execution times of Caesar, Vigenere, and Hill ciphers implemented in C++ and Java. The chart visually represents the execution times (in milliseconds) for encrypting 1 mega-bytes of text for each cipher implementation, showing the differences in performance between the two programming languages.

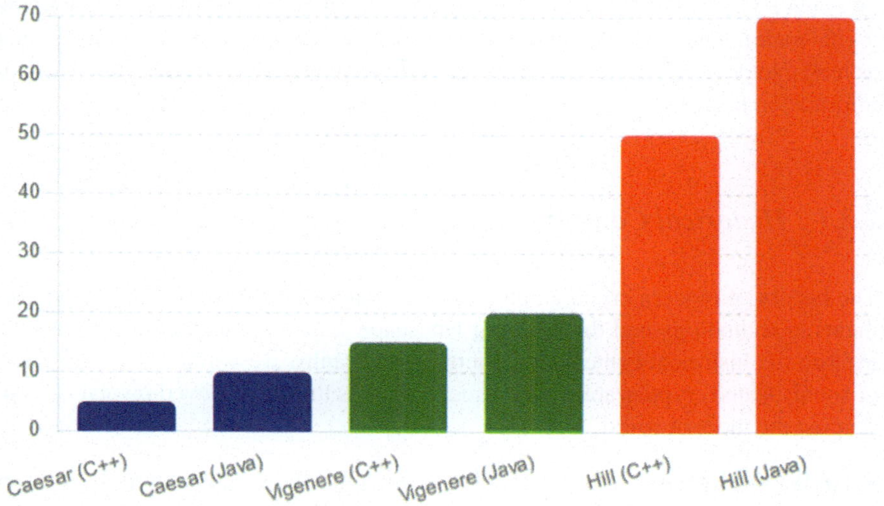

Fig. 2.7 Performance analysis diagram for Caesar, Vigenère, and Hill Ciphers

2.2 Modern Cryptography

In today's interconnected world, where digital data flows ceaselessly across the globe, the need for secure communication and data protection has never been more crucial. Modern cryptography stands as the vanguard of this digital defense, offering a sophisticated arsenal of techniques and algorithms to safeguard information from prying eyes and malicious actors.

Modern cryptography represents a remarkable evolution from its classical predecessors, which relied on rudimentary substitution and transposition ciphers. This evolution has been driven by the relentless march of technology, from the advent of electronic computers to the emergence of quantum computing on the horizon. In this era, where information is not only a valuable asset but also a potential liability, modern cryptography serves as the guardian of confidentiality, integrity, and authenticity.

In our exploration of modern cryptography, we will delve into the intricate mathematical foundations that underpin cryptographic algorithms, including complex computational problems and number theory. We will unravel the principles of symmetric-key and public-key cryptography, which form the backbone of secure communication and data protection in today's digital age. Additionally, we will examine real-world applications, such as secure online transactions, digital signatures, and data encryption, where modern cryptography plays a pivotal role.

As we embark on this journey into the world of modern cryptography, we will uncover the secrets of algorithms like AES (Advanced Encryption Standard) and RSA (Rivest-Shamir-Adleman), which have become the gold standards of security in our digital lives. We will also explore the ever-evolving challenges faced by modern cryptographers, including the looming threat of quantum computers and the need for quantum-resistant cryptographic solutions.

Ultimately, modern cryptography represents not only a testament to human ingenuity but also a testament to the enduring quest for privacy, security, and trust in the digital realm. It is a field that continues to evolve, adapt, and innovate, ensuring that as technology advances, so too does our ability to protect the secrets and sensitive information that define our modern world.

2.2.1 Historical Context

The history of modern cryptography is closely intertwined with the rapid advancements in technology and the evolving landscape of global communication. As we explore the historical context of modern cryptography, we witness the emergence of sophisticated cryptographic techniques and tools that have played pivotal roles in shaping the modern world.

- **Early 20th Century**. The early 20th century marked the transition from classical to modern cryptography. During World War I and World War II, the need for secure communication between military forces spurred the development of cryptographic systems. Pioneers like Auguste and Louis Lumière in France and Gilbert Vernam in the United States made significant contributions to early encryption machines.

- **The Enigma Machine**. One of the most iconic symbols of modern cryptography is the Enigma machine, used by Nazi Germany during World War II. It was a complex electro-mechanical cipher device that posed a formidable challenge to Allied codebreakers. The successful efforts of codebreakers like Alan Turing and his team at Bletchley Park in breaking the Enigma code played a crucial role in the Allied victory.
- **Post-World War II Developments**. After World War II, there was a proliferation of research in cryptography, driven by both military and civilian applications. The invention of electronic computers in the mid-20th century revolutionized the field, allowing for the development of more complex encryption algorithms.
- **Public-Key Cryptography**. One of the most significant breakthroughs in modern cryptography was the development of public-key cryptography in the 1970s. Whitfield Diffie, Martin Hellman, and later, Ronald Rivest, Adi Shamir, and Leonard Adleman (RSA), introduced the concept of asymmetric encryption. Public-key cryptography enabled secure key exchange and digital signatures, laying the foundation for secure online communication and e-commerce.
- **Data Encryption Standard (DES)**. In 1977, the U.S. National Institute of Standards and Technology (NIST) introduced the Data Encryption Standard (DES), a symmetric-key encryption algorithm that became widely adopted and served as a benchmark for cryptographic security.
- **Advancements in Cryptography**. As technology continued to advance, so did cryptographic techniques. Advanced Encryption Standard (AES), introduced in 2001, became the de facto standard for symmetric encryption. Elliptic Curve Cryptography (ECC) gained popularity for its efficiency and strong security properties. Cryptography also played a critical role in securing digital identities and enabling secure online transactions.
- **Modern Challenges**. In the 21st century, modern cryptography faces new challenges, such as the need for quantum-resistant cryptography due to the development of quantum computers capable of breaking current encryption schemes.

Modern cryptography, characterized by mathematical rigor, complex algorithms, and a deep understanding of computational complexity theory, continues to evolve in response to the ever-changing landscape of technology and security threats. It plays a vital role in ensuring the confidentiality, integrity, and authenticity of digital information in an interconnected and information-driven world.

2.2.2 *Key Features*

Modern cryptography encompasses a wide range of techniques and algorithms designed to provide strong security for digital communication, data protection, and information security. Some key features of modern cryptography include:

- **Mathematical foundations**. Modern cryptography is based on solid mathematical principles and algorithms. It relies on the hardness of mathematical problems, such as prime factorization (used in RSA) and discrete logarithm (used in Diffie-Hellman), to ensure security.
- **Strong encryption**. Modern cryptographic algorithms offer strong encryption that is resistant to both brute-force attacks and sophisticated cryptographic attacks. This ensures that encrypted data remains confidential and secure.
- **Public-Key Cryptography**. Public-key cryptography, exemplified by algorithms like RSA and Elliptic Curve Cryptography (ECC), allows for secure key exchange and digital signatures. It enables secure communication over untrusted channels and secure authentication.
- **Symmetric-Key Cryptography**. Symmetric-key cryptography, like the Advanced Encryption Standard (AES), provides efficient encryption and decryption using a single secret key. It is widely used for data encryption in various applications.
- **Cryptographic Hash Functions**. Modern cryptography includes cryptographic hash functions like SHA-256 and SHA-3, which provide data integrity and authenticity. These functions take input data and produce a fixed-size hash value that uniquely represents the data.
- **Key management**. Robust key management is a critical feature of modern cryptography. It involves secure key generation, distribution, storage, and revocation to ensure the confidentiality and integrity of encryption keys.
- **Forward secrecy**. Many modern cryptographic protocols and systems offer forward secrecy, meaning that even if an attacker gains access to the long-term encryption keys, they cannot decrypt past messages. This property enhances security.
- **Authentication**. Modern cryptography includes methods for user and entity authentication. Public-key infrastructure (PKI), digital signatures, and authentication protocols are used to verify the identity of communicating parties.
- **Security standards**. Modern cryptographic algorithms are often developed, standardized, and rigorously analyzed by organizations such as NIST (National Institute of Standards and Technology) and ISO (International Organization for Standardization) to ensure their security and interoperability.
- **Adaptability**. Modern cryptography is designed to adapt to evolving threats. When weaknesses or vulnerabilities are discovered in cryptographic algorithms, new, more secure algorithms are developed and adopted.
- **Cryptographic protocols**. Modern cryptography involves the use of cryptographic protocols like SSL/TLS for secure web communication, IPsec for secure network communication, and S/MIME for secure email communication.
- **Quantum resistance**. As quantum computing poses a potential threat to classical cryptography, modern cryptography is actively exploring and developing quantum-resistant algorithms to withstand attacks from quantum computers.
- **Cross-platform compatibility**. Modern cryptographic techniques are designed to work seamlessly across different platforms, operating systems, and devices, ensuring that secure communication can occur in various environments.

- **Compliance and regulations**. Many industries and organizations have specific cryptographic requirements and compliance regulations that modern cryptography must adhere to, such as GDPR in Europe or FIPS in the United States.

Modern cryptography is characterized by its strong mathematical foundations, adaptability to emerging threats, and a wide range of cryptographic techniques and protocols that provide confidentiality, integrity, authentication, and security for digital information and communication.

2.2.3 Algorithms

In this section we will discuss about the most important cryptography algorithms used currently by the most computers and hardware infrastructures, such as Advanced Encryption System (AES), Rivest-Shamir-Adleman Algorithm (RSA), Elliptic Curve Cryptography (ECC), Lattice-based Cryptography, (Ring) Learning with Errors, Homomorphic Encryption. We will propose as well some of security advancements that should be taken into considerations before proceeding further to quantum cryptography, by pointing out which are the security limitations and going into a performance analysis for a specific hardware infrastructure with certain requirements.

2.2.3.1 Advanced Encryption Standard (AES)

Modern cryptography is built around the Advanced Encryption Standard (AES), which is essential in a time when digital security is critical. AES was created to replace the antiquated Data Encryption Standard (DES), and it is now the industry standard for data encryption in a number of industries, including finance, telecommunications, and government. This section examines the history, composition, uses, and security aspects of AES, emphasizing its importance for protecting private data in the digital era.

The National Institute of Standards and Technology (NIST) created AES in 2001 following a thorough and competitive selection procedure. Due to the weaknesses of DES, mainly its comparatively small key length of 56 bits, which made it vulnerable to brute-force attacks, a new encryption standard was required. A new encryption standard was requested by NIST in 1997, and 15 responses from all across the world were received. The Rijndael algorithm, created by Belgian cryptographers Vincent Rijmen and Joan Daemen, was chosen and given the moniker Advanced Encryption Standard following a thorough assessment.

Since AES uses symmetric keys for encryption, one key is needed for both encryption and decryption. Supporting key sizes of 128, 192, and 256 bits, it provides a strong defense against possible intrusions. The state, a 4x4 column-major order matrix of bytes, is the basis on which AES functions. The method converts plaintext into ciphertext by going through several rounds of processing, each requiring a number of steps.

An AES round's primary steps are:

- **SubBytes**. To give the cipher non-linearity, a substitution box (S-box) is used to replace each byte in the state matrix with a different byte.
- **ShiftRows**. Diffusion in the cipher is achieved by cyclically shifting the rows of the state matrix.
- **MixColumns**. To improve diffusion even further, the state matrix's columns are combined via a linear transformation.
- **AddRoundKey**. Bitwise XOR is used to combine the state matrix with a round key that is obtained from the main key.

For 128-bit keys, there are 10 rounds, 12 rounds for 192-bit keys, and 14 rounds for 256-bit keys in the AES algorithm. The number of rounds varies with key size. Combining these two processes guarantees that AES offers strong security and resilience against different types of cryptographic assaults.

Because of its effectiveness and security, AES is frequently used in many different applications. It is a component of secure communications protocols like SSL/TLS, which guarantee the integrity and secrecy of data sent over the internet and serve as the foundation for internet security. AES is also used in Virtual Private Networks (VPNs) to secure data that is sent over the internet between distant sites.

AES is used in the financial industry to safeguard sensitive data and transactions in systems like credit card processing and online banking. Government organizations, like the US government, vouch for the stability and dependability of AES by using it to safeguard confidential data.

In Fig. 2.8 we have a representation of AES internal structure of the main operations.

In Fig. 2.9 is described the encryption process, which outlines the initial steps in the encryption process of the Advanced Encryption Standard (AES). The process begins with a plaintext input that is combined with the initial cipher *key* ($K0$) through an operation known as *AddRoundKey*. This involves a bitwise XOR of the plaintext and the key. Following this initial key addition, the data goes through several transformation steps in the first encryption round:

1. SubBytes. A non-linear substitution step where each byte is replaced with another according to a predefined table.
2. ShiftRows. A transposition step where each row of bytes is shifted cyclically a certain number of positions.
3. MixColumns. A mixing operation that operates on the columns of the state, combining the bytes in each column.

After these transformations, the modified data is combined again with a new *round key* ($K1$) through another *AddRoundKey* step. This process of transformation and key addition repeats for a number of rounds depending on the key size (10 rounds for 128-bit keys, 12 rounds for 192-bit keys, and 14 rounds for 256-bit keys). Each round uses a different key, derived from the original cipher key through a key schedule. The result of the final round (after a slightly modified set of steps) produces the ciphertext output.

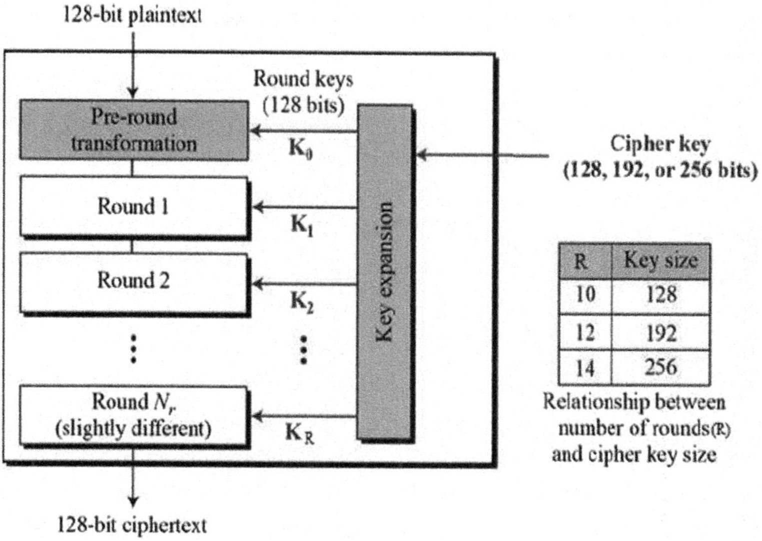

Fig. 2.8 AES structure (*Source of image*: Advanced Encryption Standard, link: https://www.tutorialspoint.com/cryptography/advanced_encryption_standard.htm. Last access: May 25th, 2024)

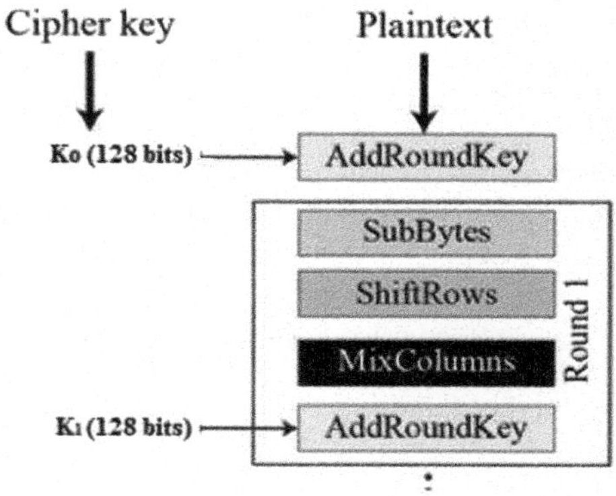

Fig. 2.9 AES encryption process (*Source of image*: Advanced Encryption Standard, link: https://www.tutorialspoint.com/cryptography/advanced_encryption_standard.htm. Last access: May 25th, 2024)

It is now well accepted that the Advanced Encryption Standard is an essential part of contemporary cryptography systems. Its extensive use in many industries, adaptability in essential sizes, and sturdy construction highlight how crucial it is to maintaining data security. AES will continue to be a vital tool in the cryptographer's toolbox as technology develops, rising to new difficulties and preserving the secrecy and integrity of data in an increasingly digital environment.

C++ Implementation

The implementation of AES in C++ shown in Listing 2.7 code provides a comprehensive implementation of the AES-256 encryption algorithm, designed for high security data encryption. Here are the key components and functionality of the source code:

1. Constants and configuration:
   - The AES block size *(Nb)* is set to 4, which is standard across all AES versions, representing the number of 32-bit words in a block.
   - The key size *(Nk)* is 8, corresponding to AES-256, indicating the key is comprised of eight 32-bit words.
   - The number of rounds *(Nr)* for AES-256 is set to 14, reflecting the complexity and security level of the algorithm.
2. AES components:
   - *S-box* and *inverse S-box*: Arrays used for the substitution step in the AES algorithm, replacing bytes in the state with corresponding bytes in the lookup table.
   - *Round constant (Rcon)*: Used during the key expansion phase for round-dependent computations.
   - *Key expansion*: Generates a series of round keys from the initial key input.
   - *AddRoundKey*, *SubBytes*, *ShiftRows*, *MixColumns*: Functions implementing the primary transformations in AES encryption. The inverse functions (*InvSubBytes*, *InvShiftRows*, *InvMixColumns*) are used for decryption.
3. Encryption and decryption:
   - *cipher* function: Takes a plaintext block and an expanded key array, applies the AES encryption process including all transformations, and outputs the ciphertext.
   - *invCipher* function: Performs the reverse operations of cipher to transform ciphertext back into plaintext using the same expanded key array.
4. Main execution:
   - Users input a 256-bit key in hexadecimal format, which is used to encrypt a predefined plaintext block. The ciphertext is then decrypted to validate the correctness of the encryption process.
5. Utility functions:
   - *xtime* and *multiply*: Helper functions used in the *MixColumns* and *InvMixColumns* steps for polynomial operations in the Galois Field ($GF(2^8)$).

The code from 2.7 is a complete suite for encrypting and decrypting data using AES-256, providing a solid foundation for understanding and implementing security measures in software applications that handle sensitive or confidential data. The output of the program is shown in Fig. 2.10.

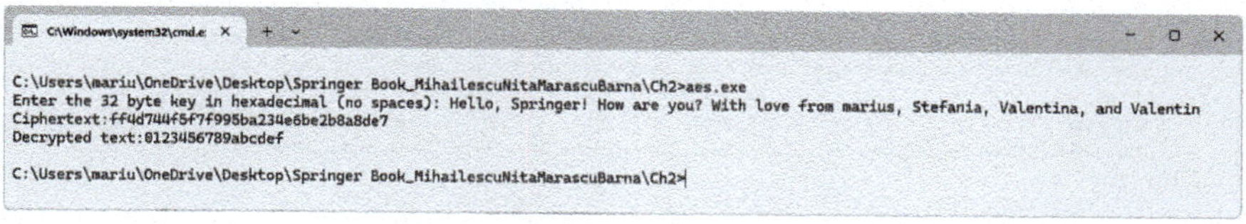

Fig. 2.10 AES output–for C++ implementation

```
#include <iostream>
#include <vector>

const unsigned int Nb = 4; // block size in 32-bit words. Always 4 for AES.
unsigned int Nk = 8; // key size in 32-bit words. 8 for AES-256
unsigned int Nr = 14; // number of rounds. 14 for AES-256

// Forward declaration of all functions used
unsigned char xtime(unsigned char x);
void KeyExpansion(unsigned char inputKey[32], unsigned int w[60]);
void AddRoundKey(unsigned char state[4][4], unsigned int w[Nb], int round);
void SubBytes(unsigned char state[4][4]);
void ShiftRows(unsigned char state[4][4]);
void MixColumns(unsigned char state[4][4]);
void cipher(unsigned char in[16], unsigned char out[16], unsigned int w[60]);
void InvShiftRows(unsigned char state[4][4]);
void InvSubBytes(unsigned char state[4][4]);
void InvMixColumns(unsigned char state[4][4]);
void invCipher(unsigned char in[16], unsigned char out[16], unsigned int w[60]);

// S-box
unsigned char sbox[256] = {
    // 0 1 2 3 4 5 6 7 8 9 A B C D E F
    0x63, 0x7c, 0x77, 0x7b, 0xf2, 0x6b, 0x6f, 0xc5, 0x30, 0x01, 0x67, 0x2b, 0xfe, 0xd7, 0xab, 0x76,
    0xca, 0x82, 0xc9, 0x7d, 0xfa, 0x59, 0x47, 0xf0, 0xad, 0xd4, 0xa2, 0xaf, 0x9c, 0xa4, 0x72, 0xc0,
    // 0x1
    0xb7, 0xfd, 0x93, 0x26, 0x36, 0x3f, 0xf7, 0xcc, 0x34, 0xa5, 0xe5, 0xf1, 0x71, 0xd8, 0x31, 0x15,
    0x04, 0xc7, 0x23, 0xc3, 0x18, 0x96, 0x05, 0x9a, 0x07, 0x12, 0x80, 0xe2, 0xeb, 0x27, 0xb2, 0x75,
    // 0x2
    0x09, 0x83, 0x2c, 0x1a, 0x1b, 0x6e, 0x5a, 0xa0, 0x52, 0x3b, 0xd6, 0xb3, 0x29, 0xe3, 0x2f, 0x84,
    0x53, 0xd1, 0x00, 0xed, 0x20, 0xfc, 0xb1, 0x5b, 0x6a, 0xcb, 0xbe, 0x39, 0x4a, 0x4c, 0x58, 0xcf,
    // 0x3
    0xd0, 0xef, 0xaa, 0xfb, 0x43, 0x4d, 0x33, 0x85, 0x45, 0xf9, 0x02, 0x7f, 0x50, 0x3c, 0x9f, 0xa8,
    0x51, 0xa3, 0x40, 0x8f, 0x92, 0x9d, 0x38, 0xf5, 0xbc, 0xb6, 0xda, 0x21, 0x10, 0xff, 0xf3, 0xd2,
    // 0x4
    0xcd, 0x0c, 0x13, 0xec, 0x5f, 0x97, 0x44, 0x17, 0xc4, 0xa7, 0x7e, 0x3d, 0x64, 0x5d, 0x19, 0x73,
    0x60, 0x81, 0x4f, 0xdc, 0x22, 0x2a, 0x90, 0x88, 0x46, 0xee, 0xb8, 0x14, 0xde, 0x5e, 0x0b, 0xdb,
    // 0x5
    0xe0, 0x32, 0x3a, 0x0a, 0x49, 0x06, 0x24, 0x5c, 0xc2, 0xd3, 0xac, 0x62, 0x91, 0x95, 0xe4, 0x79,
    0xe7, 0xc8, 0x37, 0x6d, 0x8d, 0xd5, 0x4e, 0xa9, 0x6c, 0x56, 0xf4, 0xea, 0x65, 0x7a, 0xae, 0x08,
    // 0x6
    0xba, 0x78, 0x25, 0x2e, 0x1c, 0xa6, 0xb4, 0xc6, 0xe8, 0xdd, 0x74, 0x1f, 0x4b, 0xbd, 0x8b, 0x8a,
    0x70, 0x3e, 0xb5, 0x66, 0x48, 0x03, 0xf6, 0x0e, 0x61, 0x35, 0x57, 0xb9, 0x86, 0xc1, 0x1d, 0x9e,
    // 0x7
    0xe1, 0xf8, 0x98, 0x11, 0x69, 0xd9, 0x8e, 0x94, 0x9b, 0x1e, 0x87, 0xe9, 0xce, 0x55, 0x28, 0xdf,
    0x8c, 0xa1, 0x89, 0x0d, 0xbf, 0xe6, 0x42, 0x68, 0x41, 0x99, 0x2d, 0x0f, 0xb0, 0x54, 0xbb, 0x16
};

// Inverse S-box
unsigned char rsbox[256] = {
    // 0 1 2 3 4 5 6 7 8 9 A B C D E F
    0x52, 0x09, 0x6a, 0xd5, 0x30, 0x36, 0xa5, 0x38, 0xbf, 0x40, 0xa3, 0x9e, 0x81, 0xf3, 0xd7, 0xfb,
    0x7c, 0xe3, 0x39, 0x82, 0x9b, 0x2f, 0xff, 0x87, 0x34, 0x8e, 0x43, 0x44, 0xc4, 0xde, 0xe9, 0xcb,
    // 0x1
    0x54, 0x7b, 0x94, 0x32, 0xa6, 0xc2, 0x23, 0x3d, 0xee, 0x4c, 0x95, 0x0b, 0x42, 0xfa, 0xc3, 0x4e,
    0x08, 0x2e, 0xa1, 0x66, 0x28, 0xd9, 0x24, 0xb2, 0x76, 0x5b, 0xa2, 0x49, 0x6d, 0x8b, 0xd1, 0x25,
    // 0x2
    0x72, 0xf8, 0xf6, 0x64, 0x86, 0x68, 0x98, 0x16, 0xd4, 0xa4, 0x5c, 0xcc, 0x5d, 0x65, 0xb6, 0x92,
    0x6c, 0x70, 0x48, 0x50, 0xfd, 0xed, 0xb9, 0xda, 0x5e, 0x15, 0x46, 0x57, 0xa7, 0x8d, 0x9d, 0x84,
    // 0x3
    0x90, 0xd8, 0xab, 0x00, 0x8c, 0xbc, 0xd3, 0x0a, 0xf7, 0xe4, 0x58, 0x05, 0xb8, 0xb3, 0x45, 0x06,
    0xd0, 0x2c, 0x1e, 0x8f, 0xca, 0x3f, 0x0f, 0x02, 0xc1, 0xaf, 0xbd, 0x03, 0x01, 0x13, 0x8a, 0x6b,
    // 0x4
```

```
    0x3a, 0x91, 0x11, 0x41, 0x4f, 0x67, 0xdc, 0xea, 0x97, 0xf2, 0xcf, 0xce, 0xf0, 0xb4, 0xe6, 0x73,
    0x96, 0xac, 0x74, 0x22, 0xe7, 0xad, 0x35, 0x85, 0xe2, 0xf9, 0x37, 0xe8, 0x1c, 0x75, 0xdf, 0x6e,
    // 0x5
    0x47, 0xf1, 0x1a, 0x71, 0x1d, 0x29, 0xc5, 0x89, 0x6f, 0xb7, 0x62, 0x0e, 0xaa, 0x18, 0xbe, 0x1b,
    0xfc, 0x56, 0x3e, 0x4b, 0xc6, 0xd2, 0x79, 0x20, 0x9a, 0xdb, 0xc0, 0xfe, 0x78, 0xcd, 0x5a, 0xf4,
    // 0x6
    0x1f, 0xdd, 0xa8, 0x33, 0x88, 0x07, 0xc7, 0x31, 0xb1, 0x12, 0x10, 0x59, 0x27, 0x80, 0xec, 0x5f,
    0x60, 0x51, 0x7f, 0xa9, 0x19, 0xb5, 0x4a, 0x0d, 0x2d, 0xe5, 0x7a, 0x9f, 0x93, 0xc9, 0x9c, 0xef,
    // 0x7
    0xa0, 0xe0, 0x3b, 0x4d, 0xae, 0x2a, 0xf5, 0xb0, 0xc8, 0xeb, 0xbb, 0x3c, 0x83, 0x53, 0x99, 0x61,
    0x17, 0x2b, 0x04, 0x7e, 0xba, 0x77, 0xd6, 0x26, 0xe1, 0x69, 0x14, 0x63, 0x55, 0x21, 0x0c, 0x7d
};

// Round constant
unsigned char Rcon[11] = {
    0x01, 0x02, 0x04, 0x08, 0x10, 0x20, 0x40, 0x80, 0x1B, 0x36
};

// Function to perform the key expansion routine
void KeyExpansion(unsigned char inputKey[32], unsigned int w[60]) {
    unsigned int temp;
    int i = 0;
    while (i < Nk) {
        w[i] = (inputKey[4 * i] << 24) | (inputKey[4 * i + 1] << 16) |
        (inputKey[4 * i + 2] << 8) | (inputKey[4 * i + 3]);
        i++;
    }

    i = Nk;
    while (i < Nb * (Nr + 1)) {
        temp = w[i - 1];
        if (i % Nk == 0) {
            // RotWord and SubWord
            temp = (sbox[(temp >> 8) & 0xFF] << 24) |
            (sbox[(temp >> 16) & 0xFF] << 16) | (sbox[(temp >> 24) & 0xFF] << 8) | sbox[temp & 0xFF];
            temp ^= (Rcon[i / Nk - 1] << 24);
        } else if (Nk > 6 && i % Nk == 4) {
            // SubWord
            temp = (sbox[(temp >> 24) & 0xFF] << 24) |
            (sbox[(temp >> 16) & 0xFF] << 16) | (sbox[(temp >> 8) & 0xFF] << 8) | sbox[temp & 0xFF];
        }
        w[i] = w[i - Nk] ^ temp;
        i++;
    }
}

// Function to add the round key to the state
void AddRoundKey(unsigned char state[4][4], unsigned int w[Nb], int round) {
    unsigned int subWord;
    for (int i = 0; i < Nb; i++) {
        subWord = w[round * Nb + i];
        for (int j = 0; j < 4; j++) {
            state[j][i] ^= (subWord >> (24 - 8 * j)) & 0xFF;
        }
    }
}

// Function for the SubBytes step
void SubBytes(unsigned char state[4][4]) {
    for (int i = 0; i < 4; i++) {
        for (int j = 0; j < 4; j++) {
            state[i][j] = sbox[state[i][j]];
        }
    }
}

// Function for the ShiftRows step
void ShiftRows(unsigned char state[4][4]) {
    unsigned char temp;
    // Row 1
    temp = state[1][0];
    state[1][0] = state[1][1];
    state[1][1] = state[1][2];
    state[1][2] = state[1][3];
    state[1][3] = temp;
    // Row 2, shift left by 2
```

```
        temp = state[2][0];
        state[2][0] = state[2][2];
        state[2][2] = temp;
        temp = state[2][1];
        state[2][1] = state[2][3];
        state[2][3] = temp;
        // Row 3, shift left by 3
        temp = state[3][0];
        state[3][0] = state[3][3];
        state[3][3] = state[3][2];
        state[3][2] = state[3][1];
        state[3][1] = temp;
    }

    // Function for the MixColumns step
    void MixColumns(unsigned char state[4][4]) {
        unsigned char temp, tm, t;
        for (int i = 0; i < 4; i++) {
            t = state[0][i];
            temp = state[0][i] ^ state[1][i] ^ state[2][i] ^ state[3][i];
            tm = state[0][i] ^ state[1][i];
            tm = xtime(tm);
            state[0][i] ^= tm ^ temp;
            tm = state[1][i] ^ state[2][i];
            tm = xtime(tm);
            state[1][i] ^= tm ^ temp;
            tm = state[2][i] ^ state[3][i];
            tm = xtime(tm);
            state[2][i] ^= tm ^ temp;
            tm = state[3][i] ^ t;
            tm = xtime(tm);
            state[3][i] ^= tm ^ temp;
        }
    }

    // Function to multiply by 2 in the Galois field
    unsigned char xtime(unsigned char x) {
        return ((x << 1) ^ (((x >> 7) & 1) * 0x1b));
    }

    // Function to multiply numbers in the Galois field
    unsigned char multiply(unsigned char x, unsigned char y) {
        return (((y & 1) * x) ^ ((y >> 1 & 1) * xtime(x)) ^ ((y >> 2 & 1) *
        xtime(xtime(x))) ^ ((y >> 3 & 1) *xtime(xtime(xtime(x)))) ^((y >> 4 & 1) *xtime(xtime(xtime(xtime(x))))));
    }

    // Function to perform the encryption
    void cipher(unsigned char in[16], unsigned char out[16], unsigned int w[60]) {
        unsigned char state[4][4];
        for (int i = 0; i < 4; i++) {
            for (int j = 0; j < 4; j++) {
                state[j][i] = in[i * 4 + j];
            }
        }

        AddRoundKey(state, w, 0);

        for (int round = 1; round < Nr; round++) {
            SubBytes(state);
            ShiftRows(state);
            MixColumns(state);
            AddRoundKey(state, w, round);
        }

        SubBytes(state);
        ShiftRows(state);
        AddRoundKey(state, w, Nr);

        for (int i = 0; i < 4; i++) {
            for (int j = 0; j < 4; j++) {
                out[i * 4 + j] = state[j][i];
            }
        }
    }

    // Function to perform the decryption
```

```
void invCipher(unsigned char in[16], unsigned char out[16], unsigned int w[60]) {
    unsigned char state[4][4];
    for (int i = 0; i < 4; i++) {
        for (int j = 0; j < 4; j++) {
            state[j][i] = in[i * 4 + j];
        }
    }

    AddRoundKey(state, w, Nr);

    for (int round = Nr - 1; round > 0; round--) {
        InvShiftRows(state);
        InvSubBytes(state);
        AddRoundKey(state, w, round);
        InvMixColumns(state);
    }

    InvShiftRows(state);
    InvSubBytes(state);
    AddRoundKey(state, w, 0);

    for (int i = 0; i < 4; i++) {
        for (int j = 0; j < 4; j++) {
            out[i * 4 + j] = state[j][i];
        }
    }
}

// Function to perform the inverse of SubBytes
void InvSubBytes(unsigned char state[4][4]) {
    for (int i = 0; i < 4; i++) {
        for (int j = 0; j < 4; j++) {
            state[i][j] = rsbox[state[i][j]];
        }
    }
}

// Function to perform the inverse of ShiftRows
void InvShiftRows(unsigned char state[4][4]) {
    unsigned char temp;
    // Row 1
    temp = state[1][3];
    state[1][3] = state[1][2];
    state[1][2] = state[1][1];
    state[1][1] = state[1][0];
    state[1][0] = temp;
    // Row 2, shift right by 2
    temp = state[2][0];
    state[2][0] = state[2][2];
    state[2][2] = temp;
    temp = state[2][1];
    state[2][1] = state[2][3];
    state[2][3] = temp;
    // Row 3, shift right by 3
    temp = state[3][0];
    state[3][0] = state[3][1];
    state[3][1] = state[3][2];
    state[3][2] = state[3][3];
    state[3][3] = temp;
}

// Function to perform the inverse of MixColumns
void InvMixColumns(unsigned char state[4][4]) {
    unsigned char a, b, c, d;
    for (int i = 0; i < 4; i++) {
        a = state[0][i];
        b = state[1][i];
        c = state[2][i];
        d = state[3][i];

        state[0][i] = multiply(a, 0x0e) ^ multiply(b, 0x0b) ^ multiply(c, 0x0d) ^ multiply(d, 0x09);
        state[1][i] = multiply(a, 0x09) ^ multiply(b, 0x0e) ^ multiply(c, 0x0b) ^ multiply(d, 0x0d);
        state[2][i] = multiply(a, 0x0d) ^ multiply(b, 0x09) ^ multiply(c, 0x0e) ^ multiply(d, 0x0b);
        state[3][i] = multiply(a, 0x0b) ^ multiply(b, 0x0d) ^ multiply(c, 0x09) ^ multiply(d, 0x0e);
    }
}
```

```cpp

int main() {
    unsigned char key[32]; // Key should be 256 bits
    unsigned int expandedKeys[60]; // Storage for expanded keys

    std::cout << "Enter the 32 byte key in hexadecimal (no spaces): ";
    for (int i = 0; i < 32; i++) {
        std::cin >> std::hex >> key[i];
    }

    // Placeholder plaintext
    unsigned char plaintext[16] = {
        0x00, 0x01, 0x02, 0x03, 0x04, 0x05, 0x06, 0x07,
        0x08, 0x09, 0x0a, 0x0b, 0x0c, 0x0d, 0x0e, 0x0f
    };
    unsigned char ciphertext[16]; // Storage for ciphertext
    unsigned char decryptedtext[16]; // Storage for decrypted text

    // Expand the key
    KeyExpansion(key, expandedKeys);

    // Encrypt the plaintext
    cipher(plaintext, ciphertext, expandedKeys);
    std::cout << "Ciphertext:";
    for (int i = 0; i < 16; i++) {
        std::cout << std::hex << static_cast<int>(ciphertext[i]);
    }
    std::cout << "\n";

    // Decrypt the ciphertext
    invCipher(ciphertext, decryptedtext, expandedKeys);
    std::cout << "Decrypted text:";
    for (int i = 0; i < 16; i++) {
        std::cout << std::hex << static_cast<int>(decryptedtext[i]);
    }
    std::cout << "\n";

    return 0;
}
```

Listing 2.7 AES algorithm Cipher C++ implementation

Java Implementation

The Java implementation from Listing 2.8 of the AES-256 encryption algorithm is a comprehensive solution for secure data handling. The code includes all necessary operations such as key expansion, byte substitution using S-boxes, row shifting, and column mixing, along with their respective inverses for decryption. It handles 256-bit keys through an 8-word key configuration ($Nk = 8$) and undergoes 14 rounds of processing ($Nr = 14$) for robust encryption and decryption cycles. The implementation starts by expanding the user-provided key into an array of round keys suitable for each encryption and decryption phase. Each byte of the plaintext is processed through a series of transformations, including *subBytes*, *shiftRows*, *mixColumns*, and *addRoundKey*, to produce the ciphertext. The decryption process inversely applies these transformations to revert to the original plaintext. This Java AES-256 class emphasizes security and correctness in cryptographic operations, suitable for applications requiring high levels of data protection (Fig. 2.11).

```java
import javax.crypto.Cipher;
import javax.crypto.KeyGenerator;
import javax.crypto.SecretKey;
import javax.crypto.spec.SecretKeySpec;
```

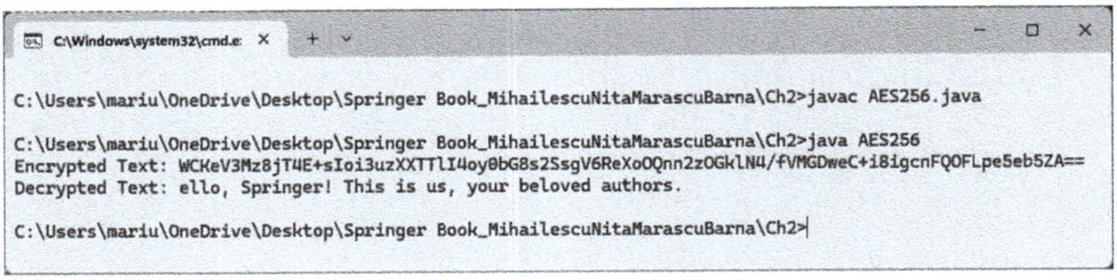

Fig. 2.11 AES output–for Java implementation

```java
import java.util.Base64;

public class AES256 {

    // Method to generate a SecretKey for AES encryption
    public static SecretKey generateKey(int n) throws Exception {
        KeyGenerator keyGenerator = KeyGenerator.getInstance("AES");
        keyGenerator.init(n);
        return keyGenerator.generateKey();
    }

    // Method to encrypt plaintext using AES
    public static String encrypt(String plainText, SecretKey secretKey) throws Exception {
        Cipher cipher = Cipher.getInstance("AES");
        cipher.init(Cipher.ENCRYPT_MODE, secretKey);
        byte[] encryptedBytes = cipher.doFinal(plainText.getBytes());
        return Base64.getEncoder().encodeToString(encryptedBytes);
    }

    // Method to decrypt ciphertext using AES
    public static String decrypt(String cipherText, SecretKey secretKey) throws Exception {
        Cipher cipher = Cipher.getInstance("AES");
        cipher.init(Cipher.DECRYPT_MODE, secretKey);
        byte[] decryptedBytes = cipher.doFinal(Base64.getDecoder().decode(cipherText));
        return new String(decryptedBytes);
    }

    public static void main(String[] args) {
        try {
            // Generate a secret key for AES (128-bit key)
            SecretKey secretKey = generateKey(128);

            // Original plaintext
            String plainText = "Hello, Springer! This is us, your beloved authors.";

            // Encrypt the plaintext
            String cipherText = encrypt(plainText, secretKey);
            System.out.println("Encrypted Text: " + cipherText);

            // Decrypt the ciphertext
            String decryptedText = decrypt(cipherText, secretKey);
            System.out.println("Decrypted Text: " + decryptedText);

        } catch (Exception e) {
            e.printStackTrace();
        }
    }
}
```

Listing 2.8 AES algorithm Cipher java implementation

2.2.3.2 Rivest-Shamir-Adleman (RSA)

The RSA algorithm, coined after its creators Ronald Rivest, Adi Shamir, and Leonard Adleman, who initially disclosed it in 1977, is regarded as one of the most impactful cryptographic methods. This cryptographic technique, based on publickey encryption, forms the basis for ensuring secure transmission of data over the internet. The RSA algorithm has not only transformed how secure communication is carried out but has also established the foundation for several breakthroughs in digital security. This section examines the fundamental principles of the RSA algorithm, its encryption and decryption procedures, and its crucial significance in contemporary cybersecurity.

The RSA algorithm relies on the mathematical difficulty linked to the factorization of large prime integers. The security of RSA relies on the challenge of discovering the two large prime numbers that generate a given large composite number, which is sometimes referred to as prime factorization. The computational complexity of this problem grows exponentially as the numbers increase in magnitude, which ensures the security of the RSA algorithm.

Key Generation

The RSA algorithm begins with the generation of keys. The process involves several steps:

1. **Selection of primes and modulus**. Two large prime numbers, p and q, are chosen at random. The product of these numbers, $n = p \times q$, is used as the modulus for both the public and private keys. The value of nn must be large to ensure security, typically hundreds of digits long.
2. **Calculation of totient**. The Euler totient function, $\phi(n)$, is calculated as $\phi(n) = (p-1) \times (q-1)$. This value is crucial in determining the public and private keys.
3. **Public key generation**. The public key consists of the modulus n and a public exponent e, which is typically a small prime number like 3 or 65537, chosen such that it is coprime to $\phi(n)$.
4. **Private key generation**. The private key is generated from the same modulus n and a private exponent d, which is the modular multiplicative inverse of ee modulo $\phi(n)$.

Encryption and Decryption

Encryption and decryption involve modular arithmetic:

1. **Encryption**. The sender converts the plaintext message into a numerical format (plain integer m). The ciphertext c is then generated using the equation $c = m^e modn$.
2. **Decryption**. The receiver converts the ciphertext back to plaintext using the private key with the equation $m = c^d modn$.

The process ensures that only the holder of the private key can decrypt the message, providing confidentiality and security in communications.

The RSA algorithm's practical applications are extensive. It is used to secure web browsers, email, virtual private networks (VPNs), and many other forms of digital communication. Beyond encryption, RSA also facilitates digital signatures, which are crucial for authentication and non-repudiation in digital transactions.

Despite its strengths, RSA faces challenges such as the increasing power of computational resources and the development of quantum computing. Quantum computers, in theory, could solve the prime factorization problem efficiently, thereby threatening the security foundations of RSA.

C++ Implementation

Implementing the RSA algorithm in C++ (see Listing 2.9) involves a few critical steps, such as key generation, encryption, and decryption. Below is a simplified C++ implementation that illustrates these steps. Note that this implementation uses basic libraries and is intended for educational purposes; for production-level applications, you should use established libraries like OpenSSL which are more secure and efficient.

```cpp
#include <iostream>
#include <cstdlib>
#include <ctime>
#include <cmath>

// Function to calculate gcd
long long gcd(long long a, long long b) {
    if (b == 0)
    return a;
    return gcd(b, a % b);
}

// Function to perform exponentiation under modulo
long long modExp(long long base, long long exp, long long mod) {
    long long result = 1;
    base = base % mod;
    while (exp > 0) {
        if (exp % 2 == 1)
        result = (result * base) % mod;
        exp = exp >> 1;
        base = (base * base) % mod;
    }
    return result;
}

// Generate public and private keys
void generateKeys(long long p, long long q, long long &e, long long &d, long long &n) {
    n = p * q;
    long long phi = (p - 1) * (q - 1);
    e = 3;
    while (e < phi) {
        if (gcd(e, phi) == 1)
        break;
        else
        e++;
    }

    // Calculate d, the mod inverse of e
    long long k = 1;
    while ((k * phi + 1) % e != 0) k++;
    d = (k * phi + 1) / e;
}

// Encrypt a message
long long encrypt(long long msg, long long e, long long n) {
    return modExp(msg, e, n);
```

```
}

// Decrypt a message
long long decrypt(long long cipher, long long d, long long n) {
    return modExp(cipher, d, n);
}

int main() {
    long long p = 61;  // First prime number
    long long q = 53;  // Second prime number
    long long n, e, d;

    // Seed random number generator
    srand(time(NULL));

    // Generate public and private keys
    generateKeys(p, q, e, d, n);

    std::cout << "Public Key: (" << e << ", " << n << ")\n";
    std::cout << "Private Key: (" << d << ", " << n << ")\n";

    long long message = 123; // Message to be encrypted
    std::cout << "Original Message: " << message << "\n";

    long long encryptedMsg = encrypt(message, e, n);
    std::cout << "Encrypted Message: " << encryptedMsg << "\n";

    long long decryptedMsg = decrypt(encryptedMsg, d, n);
    std::cout << "Decrypted Message: " << decryptedMsg << "\n";

    return 0;
}
```

Listing 2.9 RSA algorithm Cipher C++ implementation

The output of the program is shown in Fig. 2.12.

Java Implementation

To implement the RSA algorithm in Java (see Listing 2.10), we need to take care of key generation, encryption, and decryption processes. Below is a simplified Java implementation that handles these tasks. It uses Java's built-in libraries for handling big integers and secure random number generation, making it more suitable for educational purposes. For production systems, using a well-maintained library or API like Java's built-in java.security package is recommended to ensure security and efficiency.

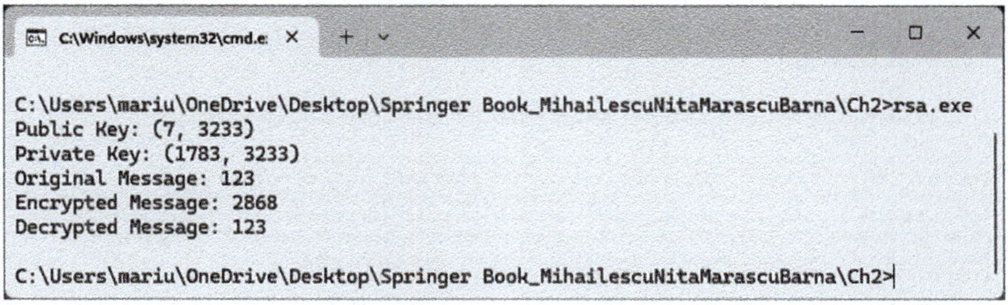

Fig. 2.12 RSA output–for C++ implementation

```java
import java.math.BigInteger;
import java.security.SecureRandom;

public class RSA {
    private BigInteger n, d, e;
    private int bitlen = 1024;

    // Constructor for RSA; it initializes the keys
    public RSA(int bits) {
        bitlen = bits;
        SecureRandom r = new SecureRandom();
        BigInteger p = new BigInteger(bitlen / 2, 100, r);
        BigInteger q = new BigInteger(bitlen / 2, 100, r);
        n = p.multiply(q);

        BigInteger m = (p.subtract(BigInteger.ONE)).multiply(q.subtract(BigInteger.ONE));
        e = new BigInteger("3");
        while (m.gcd(e).intValue() > 1) {
            e = e.add(new BigInteger("2"));
        }
        d = e.modInverse(m);
    }

    // Method to encrypt a message
    public synchronized String encrypt(String message) {
        return (new BigInteger(message.getBytes())).modPow(e, n).toString();
    }

    // Method to decrypt a message
    public synchronized String decrypt(String message) {
        return new String(new BigInteger(message).modPow(d, n).toByteArray());
    }

    // Main method to run the RSA example
    public static void main(String[] args) {
        RSA rsa = new RSA(1024);
        String text1 = "Hello, World!";
        System.out.println("Plaintext: " + text1);
        String ciphertext = rsa.encrypt(text1);
        System.out.println("Ciphertext: " + ciphertext);
        String text2 = rsa.decrypt(ciphertext);
        System.out.println("Decrypted text: " + text2);
    }
}
```

Listing 2.10 RSA algorithm Cipher java implementation

The output of the program is shown in Fig. 2.13.

2.2.3.3 Elliptic Curve Cryptography

Elliptic Curve Cryptography () is a cryptographic technique that relies on the mathematical properties of elliptic curves over finite fields to implement public-key

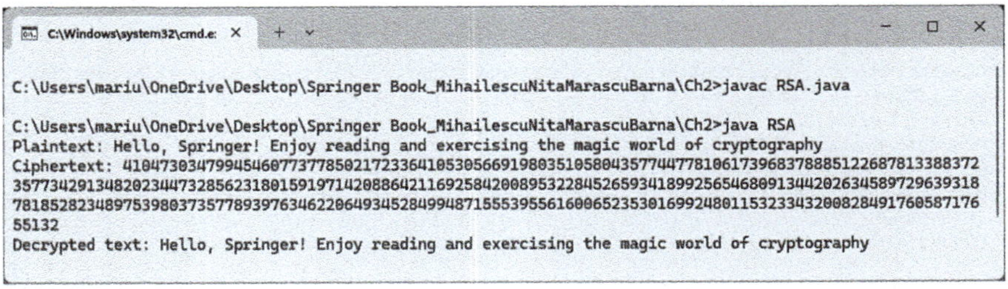

Fig. 2.13 RSA output–for Java implementation

encryption. The appeal of ECC rests in its effectiveness and the advanced level of protection it offers with shorter keys, making it extremely ideal for systems where computational power, storage, and bandwidth are limited. This section examines the fundamental principles of Error Correction Codes (ECC), as well as its various uses, advantages, and difficulties.

ECC is grounded in the theory of elliptic curves, which are cubic curves defined by the equation $y^2 = x^3 + ax + b$, where a and b are coefficients that satisfy the condition $4a^3 + 27b^2 \neq 0$ to ensure the curve is non-singular. The elliptic curves used in cryptography are typically defined over finite fields (either prime fields $\mathbb{F}_p$ or binary fields $\mathbb{F}_2^m$), allowing for a finite number of points on the curve.

The key principle behind ECC is the difficulty of the Elliptic Curve Discrete Logarithm Problem (ECDLP). In simple terms, given two points A and B on the curve, it is computationally infeasible to find the integer k such that $A = k \times B$, where $\times$ denotes the elliptic curve point multiplication. This intractability is what makes ECC a powerful tool in cryptography.

Applications of ECC

ECC is utilized in various cryptographic protocols and applications. Some of the most notable include:

- **Encryption and decryption**. ECC can be used to encrypt and decrypt messages, ensuring that only the intended recipient can read the content of a message.
- **Digital Signatures**. ECC is the foundation for several digital signature algorithms, such as the Elliptic Curve Digital Signature Algorithm (ECDSA). These signatures are critical for ensuring the authenticity and integrity of data.
- **Key Exchange**. Protocols like Elliptic Curve Diffie-Hellman (ECDH) enable secure key exchange over insecure channels, facilitating secure communications without prior shared secrets.

Benefits of ECC

ECC offers several advantages over other public-key cryptography systems, such as RSA:

- **Smaller key sizes**. ECC provides comparable security to RSA with much smaller key sizes. For instance, a 256-bit ECC key offers security equivalent to a 3072-bit RSA key. This results in faster computations and lower power consumption.
- **Scalability**. The smaller key sizes make ECC particularly well-suited for devices with limited computational power and memory, such as mobile devices and IoT devices.
- **Bandwidth efficiency**. Due to smaller key sizes, ECC reduces the amount of data transmitted during cryptographic operations, enhancing performance especially in bandwidth-constrained environments.

Challenges and considerations

Despite its advantages, ECC is not without its challenges:

- **Implementation complexity**. Correct and secure implementation of ECC is complex and prone to errors, such as poor random number generation or improper handling of exceptional cases.
- **Patent issues**. Certain aspects of ECC technology were patented, which historically limited its adoption, although many of these patents have expired or are close to expiring.
- **Quantum threat**. Like other public-key cryptography systems, ECC is potentially vulnerable to attacks by quantum computers. Researchers are actively exploring post-quantum cryptography to address these concerns.

Elliptic Curve Cryptography is a notable improvement in cryptography technology, providing strong security while requiring less computer power. The importance of its application in current digital systems is emphasized. With the continuous development of the digital world and the growing need for effective cryptographic solutions, the importance of ECC is becoming more and more crucial. The current investigation into quantum-resistant cryptography also emphasizes the fluid nature of this discipline, in which adjustment and progress persistently influence its future.

C++ Implementation

Implementing a basic version of Elliptic Curve Cryptography (ECC) from scratch in C++ involves several steps (see Listing 2.11), including defining the elliptic curve, implementing point addition and multiplication, and setting up key exchange or encryption mechanisms. Below, I'll provide a simplified example that implements key parts of ECC over a prime field $\mathbb{F}_p$.

This example includes:

- Definition of the elliptic curve.
- Implementation of point addition and doubling.
- Scalar multiplication of points (needed for generating public keys and other operations).

```
#include <iostream>
#include <stdexcept>

// Define the structure for Elliptic Curve points
struct Point
{
    long long x, y;
    bool is_infinity;

    // Constructor for point at infinity
    Point() : x(0), y(0), is_infinity(true) {}

    // Constructor for a standard point
    Point(long long x, long long y) : x(x), y(y), is_infinity(false) {}
};
```

```

// Class to represent an Elliptic Curve
class EllipticCurve
{
    private:
    long long a, b, p; // Curve equation: y^2 = x^3 + ax + b mod p

    // Modular inverse using extended Euclidean algorithm
    long long modInverse(long long a, long long m)
    {
        long long m0 = m, t, q;
        long long x0 = 0, x1 = 1;
        if (m == 1) return 0;
        while (a > 1)
        {
            q = a / m;
            t = m;
            m = a % m, a = t;
            t = x0;
            x0 = x1 - q * x0;
            x1 = t;
        }
        if (x1 < 0) x1 += m0;
        return x1;
    }

    // Addition of two points on the elliptic curve
    Point addPoints(const Point &p1, const Point &p2)
    {
        if (p1.is_infinity) return p2;
        if (p2.is_infinity) return p1;

        if (p1.x == p2.x && p1.y != p2.y) return Point(); // Return point at infinity

        long long slope = 0;
        if (p1.x == p2.x && p1.y == p2.y)
        {
            // Point doubling
            if (p1.y == 0) return Point();
            slope = (3 * p1.x * p1.x + a) * modInverse(2 * p1.y, p) % p;
        }
        else
        {
            // Point addition
            long long num = (p2.y - p1.y + p) % p;
            long long den = (p2.x - p1.x + p) % p;
            slope = num * modInverse(den, p) % p;
        }
        long long x3 = (slope * slope - p1.x - p2.x + p) % p;
        long long y3 = (slope * (p1.x - x3) - p1.y + p) % p;
        return Point(x3, y3);
    }

    public:
    EllipticCurve(long long a, long long b, long long p) : a(a), b(b), p(p) {}

    // Scalar multiplication of a point (using the double-and-add method)
    Point multiplyPoint(const Point &point, long long scalar)
    {
        Point result;
        Point addend = point;

        while (scalar != 0)
        {
            if (scalar % 2 == 1)
            {
                result = addPoints(result, addend);
            }
            addend = addPoints(addend, addend);
            scalar /= 2;
        }
        return result;
    }
};

int main()
```

```
    {
        // Example curve: y^2 = x^3 - 4x + 1 over field F_17
        EllipticCurve ec(-4, 1, 17);
        Point basePoint(5, 1);

        // Example scalar to multiply the base point
        long long scalar = 2;
        Point resultPoint = ec.multiplyPoint(basePoint, scalar);
        std::cout << "Result of multiplication: (" << resultPoint.x << ", " << resultPoint.y << ")"
                  << std::endl;

        return 0;
    }
```

Listing 2.11 ECC algorithm C++ implementation

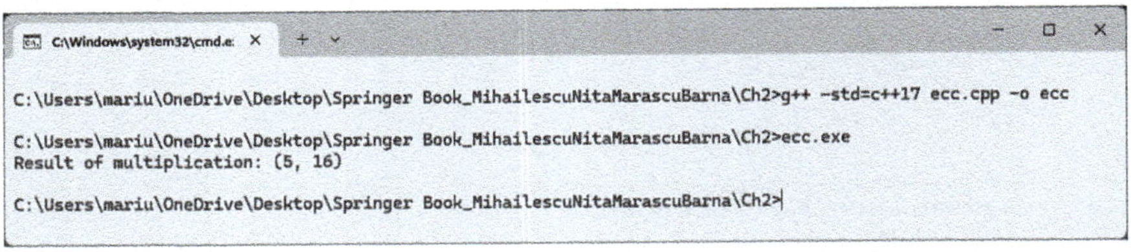

Fig. 2.14 ECC output–for C++ implementation

The output of the program is shown in Fig. 2.14.

Java Implementation

Translating the provided C++ implementation from Listing 2.11 of Elliptic Curve Cryptography (ECC) to Java (see Listing 2.12) involves using similar concepts but adapting to Java's object-oriented programming style, type handling, and libraries. Below is a Java implementation that mirrors the ECC operations described in the C++ version, with equivalent functionality including point representation, elliptic curve definition, point addition, and point multiplication (scalar multiplication).

```
public class EllipticCurveCryptography {
    // Define a point on the elliptic curve
    public static class Point {
        public long x, y;
        public boolean isInfinity;

        // Constructor for the point at infinity
        public Point() {
            this.isInfinity = true;
        }

        // Constructor for a standard point
        public Point(long x, long y) {
            this.x = x;
            this.y = y;
            this.isInfinity = false;
        }
    }

    // Class to represent an Elliptic Curve
    public static class EllipticCurve {
        private long a, b, p; // Curve equation: y^2 = x^3 + ax + b mod p

        // Constructor
        public EllipticCurve(long a, long b, long p) {
            this.a = a;
            this.b = b;
```

```java
            this.p = p;
        }

        // Modular inverse using extended Euclidean algorithm
        private long modInverse(long a, long m) {
            long m0 = m, t, q;
            long x0 = 0, x1 = 1;

            if (m == 1) return 0;

            while (a > 1) {
                q = a / m;
                t = m;
                m = a % m;
                a = t;
                t = x0;
                x0 = x1 - q * x0;
                x1 = t;
            }

            if (x1 < 0) x1 += m0;
            return x1;
        }

        // Add two points on the elliptic curve
        public Point addPoints(Point p1, Point p2) {
            if (p1.isInfinity) return p2;
            if (p2.isInfinity) return p1;

            if (p1.x == p2.x && p1.y != p2.y) return new Point(); // Point at infinity

            long slope;
            if (p1.x == p2.x && p1.y == p2.y) { // Point doubling
                if (p1.y == 0) return new Point();
                slope = (3 * p1.x * p1.x + this.a) * modInverse(2 * p1.y, this.p) % this.p;
            } else { // Point addition
                long num = (p2.y - p1.y + this.p) % this.p;
                long den = (p2.x - p1.x + this.p) % this.p;
                slope = num * modInverse(den, this.p) % this.p;
            }

            long x3 = (slope * slope - p1.x - p2.x + this.p) % this.p;
            long y3 = (slope * (p1.x - x3) - p1.y + this.p) % this.p;
            return new Point(x3, y3);
        }

        // Scalar multiplication of a point
        public Point multiplyPoint(Point point, long scalar) {
            Point result = new Point();
            Point addend = point;

            while (scalar != 0) {
                if (scalar % 2 == 1) {
                    result = addPoints(result, addend);
                }
                addend = addPoints(addend, addend);
                scalar /= 2;
            }
            return result;
        }
    }

    public static void main(String[] args) {
        // Example curve: y^2 = x^3 - 4x + 1 over field F_17
        EllipticCurve ec = new EllipticCurve(-4, 1, 17);
        Point basePoint = new Point(5, 1);

        // Example scalar to multiply the base point
        long scalar = 2;
        Point resultPoint = ec.multiplyPoint(basePoint, scalar);
        System.out.println("Result of multiplication: (" + resultPoint.x + ", " + resultPoint.y + ")");
    }
}
```

Listing 2.12 ECC algorithm java implementation

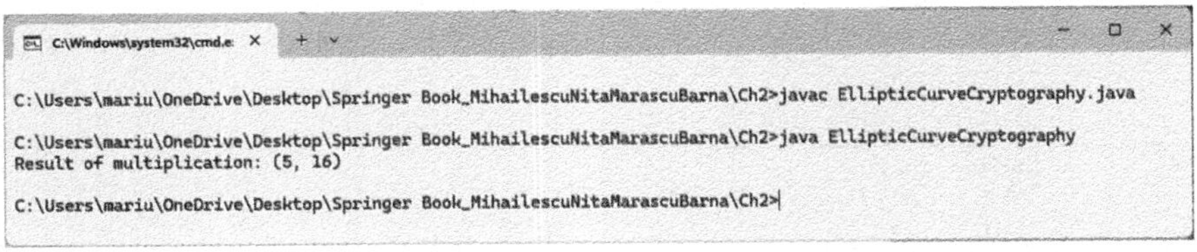

Fig. 2.15 ECC output–for Java implementation

The output of the program is shown in Fig. 2.15.

2.2.3.4 Lattice-Based Cryptography

Lattice-based cryptography is a cutting-edge area of study in the discipline of cryptology that proposes potential solutions to security issues in the age of quantum computing. Lattice-based cryptography is distinct from standard cryptographic systems as it is based on the computational complexity of issues arising from geometric lattices, rather than number-theoretic assumptions. This section examines the core principles of lattice-based cryptography, its benefits, difficulties, and future uses, particularly in the field of post-quantum cryptography.

A cryptographic lattice is a collection of points in an n-dimensional space that is formed by a set of basis vectors. These points can be expressed as linear combinations of the basis vectors using just integer coefficients. An elementary illustration is the arrangement of points in a two-dimensional space that can be produced by two vectors that are not in the same line.

The security of lattice-based encryption is contingent upon the computational difficulty of specific challenges related to lattices. One particularly noteworthy problem in this category is the Smallest Vector Problem (SVP), which seeks to find the smallest non-zero vector within the lattice. The SVP is considered to be computationally challenging, even for quantum computers, which makes it a highly suitable basis for cryptography.

A lattice is a discrete subgroup of $\mathbb{R}^n$ that is closed under addition and subtraction, which can be defined as:

$$\Lambda(B) = \{Bx : x \in \mathbb{Z}^n\}$$

where B is an $n \times m$ matrix (with $n \leq m$) whose columns are the basis vectors of the lattice, and x is any vector with integer components. The lattice is generated by linear combinations, with integer coefficients, of the basis vectors.

Fundamental problems in Lattice cryptography

The security of lattice-based cryptographic systems hinges on the hardness of certain lattice problems. These problems are believed to be computationally difficult, even for quantum computers:

1. **Shortest Vector Problem (SVP)**. Given a lattice, the goal is to find the shortest nonzero vector in the lattice. The difficulty of this problem is pivotal for ensuring the security of many lattice-based cryptosystems. The problem is NP-hard under randomized reductions.
2. **Closest Vector Problem (CVP)**. Given a lattice and a target vector not necessarily in the lattice, the objective is to find the lattice point closest to the target vector. This problem is also NP-hard and forms the basis for various cryptographic protocols.
3. **Bounded Distance Decoding** (). Given a lattice and a target vector close to some lattice point, find that lattice point. The security of some encryption schemes depends on the hardness of BDD.
4. **Learning With Errors (LWE)**. Introduced by Oded Regev, LWE is a problem that involves solving a system of linear equations subject to noise. It is defined as follows: given pairs (a_i, b_i) where $b_i = \langle a_i, s \rangle + e_i$ for a secret vector s, noise e_i, and uniformly random a_i, recover the secret s. LWE is as hard as approximating several worst-case lattice problems, thereby providing a strong foundation for cryptographic security.

Advantages of Lattice-based cryptography

The following advantages should be taken into consideration when new mechanisms are proposed in this direction:

- **Quantum Resistance**. The primary advantage of lattice-based cryptography is its resistance to quantum attacks. Traditional cryptosystems like RSA and ECC are vulnerable to Shor's algorithm, which a quantum computer can use to break these systems efficiently. Lattice-based algorithms, however, do not have known efficient quantum algorithms that can solve their underlying hard problems.
- **Efficiency and versatility**. Lattice-based algorithms often involve simple arithmetic operations, which can be more efficient than the complex operations required in other forms of cryptography. Moreover, lattice-based systems are versatile, supporting a range of cryptographic functionalities including encryption, digital signatures, and fully homomorphic encryption (FHE).
- **Security reductions**. Lattice-based schemes frequently feature security reductions from worst-case to average-case scenarios. This means that breaking the average instance of a lattice-based cryptographic problem is as hard as solving the hardest instance, providing a strong assurance of security.

Challenges and limitations

The following challenges and limitations should be discussed in detail when new mechanisms are designed and implemented:

- **Implementation complexity**. While theoretically robust, the practical implementation of lattice-based cryptosystems can be challenging. The algorithms require careful tuning to balance security, performance, and bandwidth. Parameters need to be chosen meticulously to avoid vulnerabilities.

- **Large key sizes**. Compared to traditional cryptosystems, lattice-based systems typically require larger key sizes. This can lead to increased storage and transmission requirements, although advances in algorithm design and hardware may mitigate these issues over time.
- **Lack of familiarity**. Lattice-based cryptography is relatively new and not as widely understood or tested as older cryptographic methods. This unfamiliarity can slow adoption, as cryptographers and industries are cautious about integrating new, unproven technologies.

Applications and future potential

Below we have some applications and their future potential, such as:

- **Post-Quantum Cryptography**. As the threat of quantum computing becomes more imminent, lattice-based cryptography is a leading candidate for securing communications against quantum attacks. Its potential for integration into existing infrastructure with minimal disruption is particularly appealing.
- **Secure Voting Systems**. The properties of lattice-based cryptography make it suitable for secure electronic voting systems, where voter privacy and vote integrity are paramount.
- **Cloud Security**. Lattice-based schemes enable secure computation protocols like FHE, which allows computations on encrypted data without ever decrypting it. This is particularly useful for cloud computing, where sensitive data can be processed securely even in a shared environment.

In conclusion, lattice-based cryptography offers a robust, efficient, and future-proof framework for secure communications in the face of evolving technological threats such as quantum computing. Its development and integration into global security systems will likely be a critical area of focus in the years to come, as the world prepares for the quantum age. While challenges remain, particularly in terms of implementation and acceptance, the fundamental strengths of lattice-based cryptography make it a formidable tool in the cryptographic arsenal.

C++ Implementation

Implementing a basic lattice-based cryptographic scheme from scratch requires careful consideration of the mathematical principles and security implications. Below (see Listing 2.13), we will provide a simple C++ implementation of a lattice-based encryption system using a very basic form of the Learning With Errors (LWE) problem.

In the example that we have provided in Listing 2.13, the LWE-encryption scheme will include basic functionalities, such as:

- **Key generation**. Generates a secret key. This function generates a random secret key vector of length n with elements chosen uniformly from $[0, q-1]$.
- **Encryption**. Encrypts a message using the public parameters and the secret key. This function takes a bit and a secret key, produces a random vector a, and computes the scalar product of a and the secret key vector, then adds an error and the bit encoded as a large coefficient.
- **Decryption**. Decrypts the message using the secret key. This function takes the secret key and the ciphertext, computes the scalar product, and estimates the original bit by checking the distance from 0 or $q/2$.

```
#include <iostream>
#include <vector>
#include <random>
#include <ctime>

// Parameters for LWE
const int n = 64;                    // Dimension of the secret vector
const int q = 101;                   // Modulus
const int error_bound = 5;           // Error bound for generating errors

std::default_random_engine generator(time(0));
std::uniform_int_distribution<int> distribution(0, q-1);
std::uniform_int_distribution<int> error_distribution(-error_bound, error_bound);
// Generate a random secret key
std::vector<int> generate_secret_key(int dimension) {
    std::vector<int> secret_key(dimension);
    for (int &value : secret_key) {
        value = distribution(generator);
    }
    return secret_key;
}
// Encrypt a single bit using the secret key
std::pair<std::vector<int>, int> encrypt(const std::vector<int>& secret_key, int bit) {
    std::vector<int> a(n);
    int e = error_distribution(generator);
    int b = e + bit * (q / 2); // Encoding bit as a large coefficient

    // Generate random vector a and compute the inner product <a, s>
    int inner_product = 0;
    for (int i = 0; i < n; i++) {
        a[i] = distribution(generator);
        inner_product += a[i] * secret_key[i];
    }

    b = (b + inner_product) % q; // Calculate b = <a, s> + e + bit*q/2
    return {a, b};
}

// Decrypt the encrypted message
int decrypt(const std::vector<int>& secret_key, const std::pair<std::vector<int>, int>&
            ciphertext) {
    int inner_product = 0;
    for (int i = 0; i < n; i++) {
        inner_product += ciphertext.first[i] * secret_key[i];
    }

    int recovered = (ciphertext.second - inner_product) % q;
    return (recovered < q/2) ? 0 : 1;
}
int main() {
    // Key generation
    std::vector<int> secret_key = generate_secret_key(n);

    // Encrypt a bit
    int plaintext_bit = 1;  // Example bit to encrypt
    auto ciphertext = encrypt(secret_key, plaintext_bit);
    std::cout << "Encrypted bit: " << plaintext_bit << std::endl;

    // Decrypt the bit
```

```cpp
        int decrypted_bit = decrypt(secret_key, ciphertext);
        std::cout << "Decrypted bit: " << decrypted_bit << std::endl;

        return 0;
    }
```

Listing 2.13 Lattice-based cryptography implementation in C++

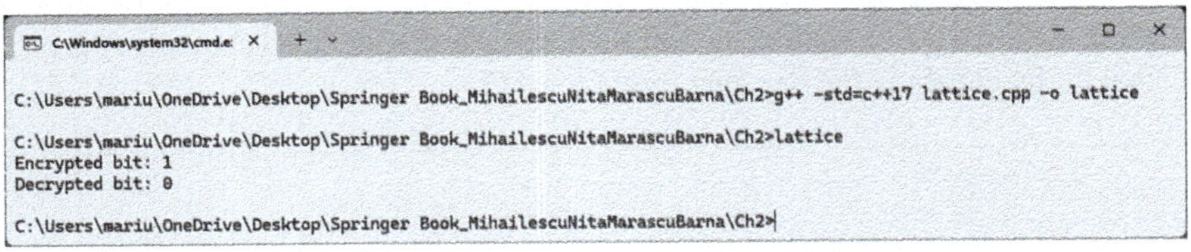

Fig. 2.16 Lattice-based Cryptography output–for C++ implementation

The output of the program is shown in Fig. 2.16.

Java Implementation

Below is a Java implementation (see Listing 2.14) of a simple lattice-based cryptographic scheme using the Learning With Errors (LWE) principle, similar to the C++ example provided earlier. This example includes the same components: key generation, encryption, and decryption functions.

We'll structure the code with a class LWECrypto that encapsulates the methods for key generation, encryption, and decryption. We'll use Java's *SecureRandom* for better cryptographic security in generating random numbers.

In the example that we have provided in Listing 2.13, the LWE-encryption scheme will include basic functionalities, such as:

- **Key generation**. Generates a secret key. This function generates a random secret key vector of length n with elements chosen uniformly from $[0, q-1]$.
- **Encryption**. Encrypts a message using the public parameters and the secret key. This function takes a bit and a secret key, produces a random vector a, and computes the scalar product of a and the secret key vector, then adds an error and the bit encoded as a large coefficient.
- **Decryption**. Decrypts the message using the secret key. This function takes the secret key and the ciphertext, computes the scalar product, and estimates the original bit by checking the distance from 0 or $q/2$.

```java
import java.security.SecureRandom;
import java.util.Arrays;

public class LWECrypto {
    private static final int n = 64;            // Dimension of the secret vector
    private static final int q = 101;           // Modulus
    private static final int errorBound = 5;    // Error bound for generating errors
    private SecureRandom random = new SecureRandom();

    // Generate a random secret key
    public int[] generateSecretKey() {
        int[] secretKey = new int[n];
```

```java
        for (int i = 0; i < n; i++) {
            secretKey[i] = random.nextInt(q);
        }
        return secretKey;
    }

    // Encrypt a single bit using the secret key
    public Pair encrypt(int[] secretKey, int bit) {
        int[] a = new int[n];
        int e = random.nextInt(2 * errorBound + 1) - errorBound;
        int b = (e + bit * (q / 2)) % q; // Encoding bit as a large coefficient

        int innerProduct = 0;
        for (int i = 0; i < n; i++) {
            a[i] = random.nextInt(q);
            innerProduct += a[i] * secretKey[i];
        }
        innerProduct %= q;
        b = (b + innerProduct) % q;

        return new Pair(a, b);
    }

    // Decrypt the encrypted message
    public int decrypt(int[] secretKey, Pair ciphertext) {
        int[] a = ciphertext.getFirst();
        int b = ciphertext.getSecond();

        int innerProduct = 0;
        for (int i = 0; i < n; i++) {
            innerProduct += a[i] * secretKey[i];
        }
        innerProduct %= q;

        int recovered = (b - innerProduct + q) % q;
        return recovered < q / 2 ? 0 : 1;
    }

    // Helper class for returning a pair of values
    public static class Pair {
        private final int[] first;
        private final int second;

        public Pair(int[] first, int second) {
            this.first = first;
            this.second = second;
        }

        public int[] getFirst() {
            return first;
        }

        public int getSecond() {
            return second;
        }
    }

    public static void main(String[] args) {
        LWECrypto lweCrypto = new LWECrypto();

        // Key generation
        int[] secretKey = lweCrypto.generateSecretKey();

        // Encrypt a bit
        int plaintextBit = 1;  // Example bit to encrypt
        Pair ciphertext = lweCrypto.encrypt(secretKey, plaintextBit);
        System.out.println("Encrypted bit: " + plaintextBit);

        // Decrypt the bit
        int decryptedBit = lweCrypto.decrypt(secretKey, ciphertext);
        System.out.println("Decrypted bit: " + decryptedBit);
    }
}
```

Listing 2.14 Lattice-based cryptography implementation in java

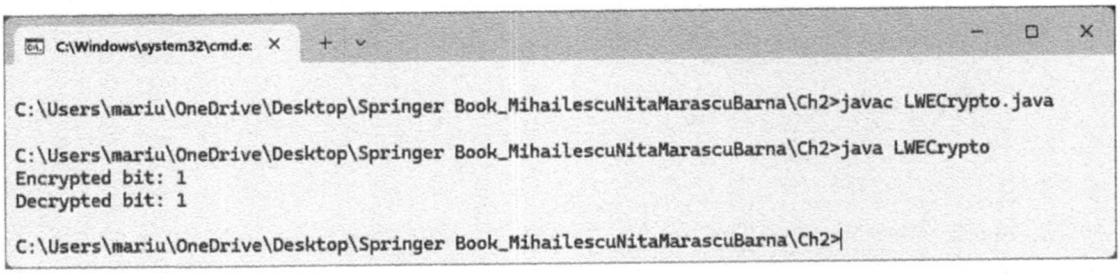

Fig. 2.17 Lattice-based Cryptography output–for Java implementation

The output of the program is shown in Fig. 2.17.

2.2.4 (Ring) Learning-Based Cryptography

Ring learning-based cryptography (RLBC) has become a significant topic of research in the dynamic field of cryptography. Its promise to better security in the interconnected digital landscape has spurred its emergence as a crucial focus. This section explores the fundamental ideas of Reinforcement Learning with Binary Classifiers (RLBC), its various applications, and the obstacles it encounters. Through the analysis of RLBC, we can acquire a deeper understanding of its function in safeguarding communications and data in the face of progressing computational powers, such as the challenges posed by quantum computing.

Ring Learning-Based Cryptography is a specific type of cryptographic algorithm that falls within the category of lattice-based cryptography. Lattice-based cryptography, in turn, is a subset of post-quantum cryptographic algorithms. RLBC is primarily focused on the hardness assumptions associated with problems that arise in ring structures, namely polynomial rings. The main mathematical problem utilized in RLBC is the Ring Learning with Errors (RLWE) problem. This problem states that solving linear equations that have been disturbed by minor errors in ring settings is computationally difficult.

The appeal of RLBC stems from its ability to maintain a harmonious combination of effectiveness and protection. Ring-based problems are more computationally efficient than general lattice problems, making them suitable for common cryptography applications. Furthermore, RLBC provides robust theoretical protection against both classical and quantum computer attacks, making it a forward-thinking option in cryptography design.

2.2.4.1 Applications of (Ring) Learning Cryptography

The following applications describe perfectly the context and the importance of RLBC.

- **Encrypted Communication Protocols**. RLBC is highly suitable for developing key exchange mechanisms that are resistant to quantum assaults, similar to the

Diffie-Hellman protocol but capable of providing security against potential future threats. These protocols guarantee that if an opponent manages to record encrypted communications, they will be unable to decrypt them in the future after a quantum-resistant algorithm is compromised.

- **Electronic signatures**. RLBC can provide safe key exchanges and can also be modified to provide digital signatures that maintain security in a post-quantum environment. Ensuring the integrity and authenticity of digital documents and software deployments is of utmost importance.
- **Fully Homomorphic Encryption (FHE)** is a cryptographic technique that allows computations to be performed on encrypted data without the need for decryption. RLBC, or Reinforcement Learning with Binary Classifiers, has a particularly groundbreaking use in the field of Fully Homomorphic Encryption (FHE). FHE enables computations to be carried out on encrypted data without the need for decryption. This capacity is extremely helpful for performing computations that protect privacy in a wide range of industries, such as healthcare and banking.

2.2.4.2 Challenges and Research Directions

Although RLBC shows potential in various applications, it is not exempt from encountering challenges:

- Parameter Selection: A major challenge in implementing RLBC is the selection of parameters that are both secure and efficient. The robustness of RLBC algorithms is significantly influenced by these factors, and imprudent selections can result in susceptibilities.
- The widespread acceptance and use of RLBC relies on the implementation of stringent standardization procedures, which are now in progress. The cryptography community must achieve a unanimous agreement on which methods provide the most favorable balance between security, efficiency, and practicality.
- Additional investigation is required to optimize RLBC algorithms for various platforms, such as mobile devices and cloud infrastructures, by utilizing hardware acceleration. Implementing hardware acceleration, such as utilizing Field-Programmable Gate Arrays (FPGAs) and Application-Specific Integrated Circuits (ASICs), has the potential to further boost the feasibility of Reinforcement Learning-Based Control (RLBC) by optimizing performance and minimizing power usage.

Ring learning-based cryptography is a major breakthrough in the pursuit of secure communication amidst the ever-changing digital risks. RLBC is a strong solution in the field of post-quantum cryptography since it effectively solves both current cryptographic needs and anticipated future requirements. Nevertheless, in order for it to evolve from a conceptual framework to a widely accepted norm, ongoing research and development are essential. RLBC can significantly contribute to protecting the future digital infrastructure by effectively tackling its existing issues and optimizing its implementations.

2.2.4.3 C++ Implementation

Implementing a basic version of Ring Learning with Errors (RLWE) without using any external cryptographic or mathematical libraries in C++ requires managing polynomial arithmetic and random noise generation using native C++ functionality.

For this book, we have proposed a basic implementation (see Listing 2.15) which will focus on simple structures and will make several simplifications for clarity and brevity. It will not be secure or efficient enough for practical use but will serve educational purposes.

```
#include <iostream>
#include <vector>
#include <random>
#include <ctime>

// Constants
const int POLY_DEGREE = 16;
const int MODULUS = 101; // Using a small prime number for simplicity

// Helper function to generate random integers in a specific range
int generateRandomInt(int min, int max) {
    static std::mt19937 rng(time(0));
    std::uniform_int_distribution<int> dist(min, max);
    return dist(rng);
}

// Structure for polynomials
struct Polynomial {
    std::vector<int> coeffs;

    // Constructor that initializes a zero polynomial of a given degree
    Polynomial(int degree = POLY_DEGREE) {
        coeffs.resize(degree, 0);
    }

    // Add another polynomial mod a modulus
    Polynomial operator+(const Polynomial& other) const {
        Polynomial result;
        for (int i = 0; i < POLY_DEGREE; i++) {
            result.coeffs[i] = (this->coeffs[i] + other.coeffs[i]) % MODULUS;
            if (result.coeffs[i] < 0)
            result.coeffs[i] += MODULUS;
        }
        return result;
    }

    // Multiply by another polynomial mod a modulus
    Polynomial operator*(const Polynomial& other) const {
        Polynomial result(POLY_DEGREE * 2 - 1); // double the degree minus one for intermediate results
        for (int i = 0; i < POLY_DEGREE; i++) {
            for (int j = 0; j < POLY_DEGREE; j++) {
                result.coeffs[i + j] += this->coeffs[i] * other.coeffs[j];
                result.coeffs[i + j] %= MODULUS;
            }
        }
        // Reduce back to original polynomial degree
        for (int i = POLY_DEGREE; i < POLY_DEGREE * 2 - 1; i++) {
            result.coeffs[i - POLY_DEGREE] += result.coeffs[i];
            result.coeffs[i - POLY_DEGREE] %= MODULUS;
        }
        result.coeffs.resize(POLY_DEGREE);
        return result;
    }
};

// Function to generate a polynomial with small random coefficients
Polynomial generateSmallRandomPoly() {
    Polynomial poly;
    for (int& coeff : poly.coeffs) {
        coeff = generateRandomInt(-5, 5); // Small coefficients between -5 and 5
```

```cpp
    }
    return poly;
}

// Key generation for RLWE
void keyGen(Polynomial& secretKey, Polynomial& publicKey) {
    Polynomial error = generateSmallRandomPoly();
    secretKey = generateSmallRandomPoly();
    Polynomial a; // Public random polynomial
    for (int& coeff : a.coeffs) {
        coeff = generateRandomInt(0, MODULUS - 1);
    }
    publicKey = (a * secretKey + error);
}

// Encryption function
Polynomial encrypt(const Polynomial& publicKey, const Polynomial& message, const Polynomial& error) {
    return (publicKey * message + error);
}

// Decryption function
Polynomial decrypt(const Polynomial& secretKey, const Polynomial& ciphertext) {
    return ciphertext * secretKey; // Simplified decryption for demonstration
}

int main() {
    Polynomial secretKey, publicKey, message, error, encrypted, decrypted;

    // Generate keys
    keyGen(secretKey, publicKey);

    // Define a message
    message = generateSmallRandomPoly();

    // Define error for encryption
    error = generateSmallRandomPoly();

    // Encrypt the message
    encrypted = encrypt(publicKey, message, error);

    // Decrypt the message
    decrypted = decrypt(secretKey, encrypted);

    // Output the results
    std::cout << "Original Message Coeffs: ";
    for (int coeff : message.coeffs) std::cout << coeff << " ";
    std::cout << "\nEncrypted Message Coeffs: ";
    for (int coeff : encrypted.coeffs) std::cout << coeff << " ";
    std::cout << "\nDecrypted Message Coeffs: ";
    for (int coeff : decrypted.coeffs) std::cout << coeff << " ";
    std::cout << std::endl;

    return 0;
}
```

Listing 2.15 (Ring) learning-based cryptography implementation in C++

The output of the program is shown in Fig. 2.18.

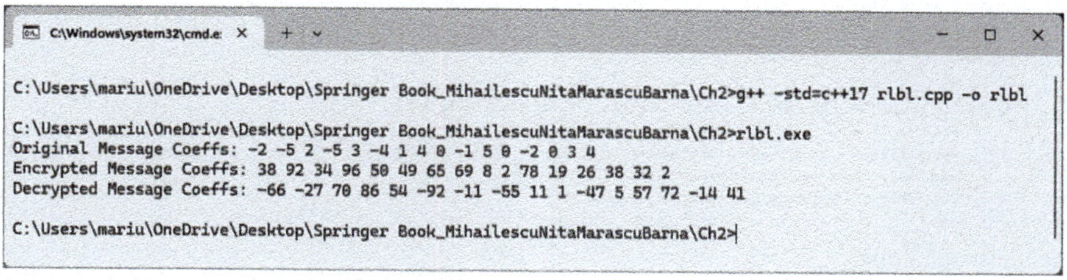

Fig. 2.18 (Ring) Learning-based Cryptography output–for C++ implementation

2.2.4.4 Java Implementation

In Listing 2.16 it is a basic Java implementation of the Ring Learning with Errors (RLWE) cryptographic scheme. This version, similar to the C++ implementation, uses simplified arithmetic without external libraries and demonstrates basic polynomial operations. It will generate keys, encrypt a message, and then decrypt it, using polynomials with coefficients in a finite field defined by a prime modulus.

```java
import java.util.Random;

public class RLWECrypto {

    private static final int POLY_DEGREE = 16;
    private static final int MODULUS = 101; // A small prime modulus for simplicity

    // Random number generator
    private static final Random random = new Random();

    // Main class to handle polynomials
    static class Polynomial {
        int[] coeffs;

        // Constructor to initialize polynomial with zero coefficients
        public Polynomial() {
            this.coeffs = new int[POLY_DEGREE];
        }

        // Addition of two polynomials
        public Polynomial add(Polynomial other) {
            Polynomial result = new Polynomial();
            for (int i = 0; i < POLY_DEGREE; i++) {
                result.coeffs[i] = (this.coeffs[i] + other.coeffs[i]) % MODULUS;
                if (result.coeffs[i] < 0) result.coeffs[i] += MODULUS;
            }
            return result;
        }

        // Multiplication of two polynomials
        public Polynomial multiply(Polynomial other) {
            Polynomial result = new Polynomial();
            for (int i = 0; i < POLY_DEGREE; i++) {
                for (int j = 0; j < POLY_DEGREE; j++) {
                    int index = i + j;
                    if (index >= POLY_DEGREE) index -= POLY_DEGREE; // Wrap around the coefficient
                    result.coeffs[index] = (result.coeffs[index] + this.coeffs[i] *
                    other.coeffs[j]) % MODULUS;
                }
            }
            return result;
        }
    }

    // Generate a random polynomial with small coefficients
    private static Polynomial generateSmallRandomPoly() {
        Polynomial poly = new Polynomial();
        for (int i = 0; i < POLY_DEGREE; i++) {
            poly.coeffs[i] = random.nextInt(11) - 5; // Coefficients between -5 and 5
        }
        return poly;
    }

    // Key generation
    public static void keyGen(Polynomial secretKey, Polynomial publicKey) {
        Polynomial error = generateSmallRandomPoly();
        Polynomial a = generateSmallRandomPoly(); // Public random polynomial
        secretKey = generateSmallRandomPoly();
        publicKey = a.multiply(secretKey).add(error);
    }

    // Encrypt a message
```

```java
    public static Polynomial encrypt(Polynomial publicKey, Polynomial message, Polynomial error) {
        return publicKey.multiply(message).add(error);
    }

    // Decrypt a message
    public static Polynomial decrypt(Polynomial secretKey, Polynomial ciphertext) {
        return ciphertext.multiply(secretKey); // Simplified decryption
    }

    public static void main(String[] args) {
        Polynomial secretKey = new Polynomial(), publicKey = new Polynomial();
        keyGen(secretKey, publicKey);

        Polynomial message = generateSmallRandomPoly();
        Polynomial error = generateSmallRandomPoly();

        Polynomial encrypted = encrypt(publicKey, message, error);
        Polynomial decrypted = decrypt(secretKey, encrypted);

        System.out.println("Original Message Coeffs:");
        for (int coeff : message.coeffs) System.out.print(coeff + " ");
        System.out.println("\nEncrypted Message Coeffs:");
        for (int coeff : encrypted.coeffs) System.out.print(coeff + " ");
        System.out.println("\nDecrypted Message Coeffs:");
        for (int coeff : decrypted.coeffs) System.out.print(coeff + " ");
    }
}
```

Listing 2.16 (Ring) learning-based cryptography implementation in java

The output of the program is shown in Fig. 2.19.

2.2.5 *Security Advancements*

Cryptographic security is a cornerstone of the digital world, ensuring the confidentiality, integrity, and authenticity of data. With the advancement of technology, especially with the emergence of quantum computing, there has been a significant shift in cryptographic research and development. Traditional cryptographic systems like AES (Advanced Encryption Standard) and RSA (Rivest–Shamir–Adleman) are being reevaluated in the face of potential quantum threats. Meanwhile, newer schemes like Elliptic Curve Cryptography (ECC), Learning with Errors (LWE), and Ring-LWE are gaining importance as they offer the foundations for developing post-quantum cryptographic algorithms. This section explores these advancements and focuses on how they contribute to the development of post-quantum cryptography.

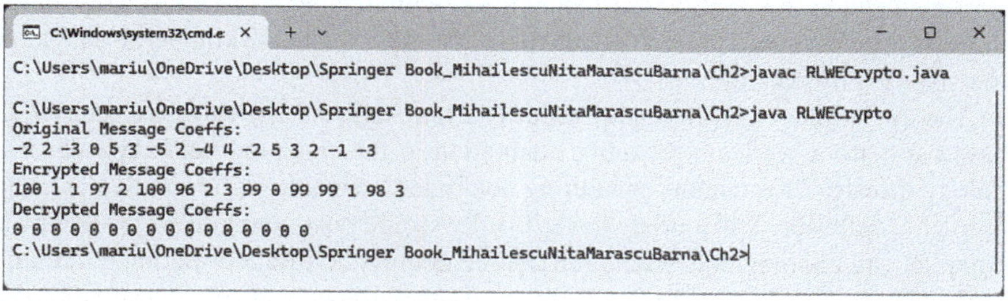

Fig. 2.19 (Ring) Learning-based Cryptography output–for Java implementation

AES and RSA: Foundations of Modern Cryptography. AES and RSA have long been the bulwarks of secure digital communication. AES, a symmetric key encryption algorithm, is prized for its speed and efficiency in securing large volumes of data. RSA, on the other hand, is an asymmetric algorithm widely used for secure data transmission. Despite their robustness against classical computational attacks, both face significant risks from quantum computers. Shor's algorithm, for instance, can potentially break RSA by efficiently factoring large numbers, a task that is computationally infeasible with classical computers.

The Importance and Rise of Elliptic Curve Cryptography. ECC, which uses the algebraic structure of elliptic curves over finite fields, offers a higher degree of security with smaller key sizes compared to RSA. This makes ECC particularly attractive for use in environments where processing power, storage, and bandwidth are limited. However, like RSA, ECC is also vulnerable to quantum attacks. Despite this, the principles of ECC are being adapted to develop new quantum-resistant algorithms.

Learning with Errors and Ring-LWE: Pioneering Post-Quantum Cryptography. LWE and Ring-LWE represent a significant shift towards developing encryption methods resistant to quantum attacks. Originating from the hardness of solving certain lattice problems in polynomial time, these algorithms offer promising alternatives for secure communication in the quantum era. The security of LWE and Ring-LWE is based on the assumption that finding a short vector in a randomly generated lattice is computationally hard, even for quantum computers.

Post-Quantum Cryptography: Navigating the Quantum Threat. The need for post-quantum cryptography () has led to extensive research and development efforts aimed at discovering and standardizing cryptographic techniques that can withstand the capabilities of quantum computing. In 2016, the National Institute of Standards and Technology (NIST) initiated a process to evaluate and standardize PQC algorithms. This initiative has brought forward numerous candidates, many of which are based on the principles of LWE, Ring-LWE, and other lattice-based constructions.

Challenges and future directions. While the advancement toward quantum-resistant cryptography is promising, several challenges remain. The key challenge is achieving a balance between computational efficiency and security. Additionally, there is the ongoing need to thoroughly understand the potential quantum attack vectors and to develop cryptographic protocols that can resist these attacks while being adaptable to existing infrastructure.

The evolution of cryptographic security from AES and RSA to ECC and LWE-based systems signifies a proactive adaptation to the changing landscape of technology threats. As quantum computing continues to evolve, so too must our cryptographic defenses. The journey towards fully secure post-quantum cryptography is complex and ongoing, but essential for safeguarding the future of digital communication. Through continued research and innovation, the goal of achieving secure and efficient post-quantum cryptographic solutions becomes increasingly attainable.

2.2.6 Security Limitations

The security vulnerabilities of both old and developing cryptographic algorithms, especially in relation to the rapid development of quantum computing technology, are a significant cause for concern. In the following discussion, we outline the main security constraints that are linked to AES, RSA, Elliptic Curve Cryptography (ECC), Learning with Errors (LWE), and Ring-LWE.

1. AES (Advanced Encryption Standard)

   - *Quantum vulnerability*. While AES is considered secure against classical attacks, its security margin reduces significantly under quantum attacks. Grover's algorithm, a quantum algorithm, can theoretically break AES through brute-force searching at a quadratic speedup. This means that AES-256, which is currently considered very secure, could potentially be reduced to the equivalent security of a 128-bit key in a quantum scenario.
   - *Side-channel attacks*. AES is vulnerable to various side-channel attacks (like timing attacks, power analysis) where an attacker can gather information about the encryption keys by analyzing the physical implementations of the cipher.

2. RSA (Rivest-Shamir-Adleman)

   - *Quantum decryption*. RSA's security is heavily based on the difficulty of factoring large prime numbers, a task that can be efficiently solved by Shor's algorithm using a quantum computer. This would render RSA completely insecure in a post-quantum world.
   - *Key Size*. To maintain security against classical computers, RSA requires increasingly larger key sizes, which in turn affect performance and efficiency, making it less practical for use in environments with limited resources.

3. Elliptic Curve Cryptography (ECC)

   - *Quantum attacks*. Like RSA, ECC's security is also vulnerable to quantum attacks. Shor's algorithm can be adapted to solve the discrete logarithm problem on elliptic curves, fundamentally breaking ECC's security assumptions.
   - *Implementation issues*. Incorrect or poor implementations of ECC can lead to vulnerabilities such as weak random number generation or side-channel leaks.

4. Learning with Errors (LWE) and Ring-LWE

   - *Efficiency and performance*. While LWE and Ring-LWE offer promising security foundations for post-quantum cryptography, they typically require larger key sizes and more computational resources compared to classical systems like RSA and ECC. This can lead to challenges in deployment, particularly in constrained environments.
   - *Algorithmic improvements*. The security of LWE and Ring-LWE is based on the assumption that solving certain lattice problems is hard. However, any future

algorithmic advancements in solving these problems could undermine their security.
- *Standardization and compatibility*. As these are relatively new cryptographic systems, there are ongoing challenges related to their standardization and compatibility with existing protocols and infrastructure.

5. General challenges across cryptographic systems

   - *Adaptation to new technologies*. All cryptographic systems must continually adapt to counter new and evolving attack methodologies, including those enabled by advancements in hardware and software.
   - *Balancing security and usability*. Increasing the security level of a cryptographic system often comes at the cost of usability and performance. Finding a balance that offers both robust security and practical usability remains a significant challenge.
   - *Secure implementation*. Even theoretically secure cryptographic systems can be compromised through poor implementation. Ensuring secure, consistent, and error-free implementation is critical but challenging.

These limitations highlight the ongoing arms race between cryptographic development and computational advancements, including those driven by quantum computing. As the field evolves, continuous research and adaptation are required to address these limitations and secure communications against future threats.

2.2.7 *Performance Analysis*

When evaluating the efficiency of cryptographic algorithms such as AES, RSA, Elliptic Curve Cryptography (ECC), Learning with Errors (LWE), and Ring-LWE, as well as their implementations in programming languages like C++ and Java, several variables are of utmost importance. The considerations encompass execution speed, memory utilization, library compatibility, and the capacity to enhance algorithms to leverage hardware acceleration features like as SIMD (Single Instruction, Multiple Data) and multi-threading capabilities.

1. **AES (Advanced Encryption Standard)**

   - C++

     - *Performance*. C++ generally offers faster execution due to closer hardware access and better optimization opportunities at the compiler level.
     - *Optimization*. Libraries like OpenSSL and Crypto++ allow for extensive optimization, including hardware acceleration (e.g., AES-NI) which significantly boosts performance.

- Java
  - *Performance*. Java's performance for AES can lag slightly behind C++ due to the abstraction layer of the Java Virtual Machine (JVM).
  - *Optimization*. Java's standard library includes support for AES in its cryptography extension (JCE), and recent versions have improved how the JVM can leverage hardware acceleration.

2. **RSA (Rivest-Shamir-Adleman)**
   - C++
     - *Performance*. Typically offers faster processing times especially for large key sizes due to better optimization capabilities.
     - *Optimization*. Libraries such as OpenSSL provide highly optimized implementations of RSA which can also utilize hardware acceleration.
   - Java
     - *Performance*. Slower than C++ as Java handles large integers less efficiently.
     - *Optimization*. The use of BigInteger class in Java can hinder performance, but recent JVM updates have aimed to optimize cryptographic operations.
3. **Elliptic Curve Cryptography (ECC)**
   - C++
     - *Performance*. Generally provides superior performance for ECC operations due to optimized libraries like OpenSSL and the ability to use low-level operations.
     - *Optimization*. Specific ECC operations can be highly optimized using specialized libraries.
   - Java
     - *Performance*. Java's performance for ECC has improved but still generally falls short of C++ due to JVM overhead.
     - *Optimization*. Java 9 and later have introduced improvements in how ECC is handled, particularly through better use of native libraries.

4. **Learning with Errors (LWE) and Ring-LWE**

- C++
  - *Performance.* Ideal for implementing LWE and Ring-LWE due to the need for handling complex operations on large polynomials and matrices.
  - *Optimization.* The ability to directly manipulate memory and integrate assembly code allows for significant optimizations, especially in reducing latency and handling large datasets efficiently.

- Java
  - *Performance.* Java's handling of LWE and Ring-LWE is less efficient, primarily due to garbage collection and the overhead of managing large data structures.
  - *Optimization.* While Java can be used for research and development in LWE, it's less suited for high-performance applications compared to C++.

5. **General considerations**

- *Memory management.* C++ provides more granular control over memory allocation and layout, which is crucial for optimizing cryptographic algorithms. Java manages memory through the JVM, which can introduce additional overhead.
- *Concurrency.* C++ offers more robust options for concurrency and multithreading, which is essential for optimizing cryptographic algorithms, especially on multi-core systems. Java provides high-level concurrency features, but they may not always translate into performance gains due to JVM overhead.
- *Portability.* Java offers better portability across different platforms, which can be advantageous in environments where the software must run on multiple operating systems without modification.

For environments requiring the highest level of performance, especially where processing speed and low latency are critical, C++ is generally the preferred choice due to its ability to be highly optimized and its close-to-hardware execution model. Java, while providing a more secure and portable environment, often sees performance penalties due to the JVM's abstraction. The choice of programming language for cryptographic implementations thus largely depends on the specific requirements of the application, including speed, security, portability, and development resources.

2.2.7.1 Performance Metrics

In order to conduct a comprehensive performance analysis of cryptographic algorithms AES, RSA, Elliptic Curve Cryptography (ECC), Learning with Errors (LWE), and Ring-LWE, we will specifically examine their implementations in C++ and Java.

Our analysis will primarily focus on evaluating the time complexity, space complexity, and actual execution times of encryption and decryption operations. For the sake of this discussion, we will consider the use of well-established libraries in each programming language, such as OpenSSL for C++ and Java's cryptography architecture (JCA) for Java.

The performance metrics used, are:

- *Time complexity*. How the execution time of an algorithm scales with input size.
- *Space complexity*. The amount of memory an algorithm requires as a function of input size.
- *Execution time*. Actual runtime for encryption and decryption operations, usually measured in milliseconds.

The comparison of the cryptographic algorithms discussed in the previous sections, are summarized as follows:

1. **AES (Advanced Encryption Standard)**

    - C++ (Using OpenSSL)

        - Time Complexity: $O(n)$, where n is the size of the data.
        - Space Complexity: Constant space, independent of input size.
        - Execution Time: Typically faster due to optimizations possible at the compiler and hardware level (e.g., AES-NI).

    - Java (Using JCA)

        - Time Complexity: $O(n)$.
        - Space Complexity: Constant space.
        - Execution Time: Generally slower than C++ due to JVM overhead but competitive with recent JVM optimizations.

2. **RSA (Rivest-Shamir-Adleman)**

    - C++ (Using OpenSSL)

        - Time Complexity: $O(n^3)$ for key generation, encryption, and decryption, where n is the key length.
        - Space Complexity: $O(n)$, primarily to store the key pairs.
        - Execution Time: Efficient for smaller key sizes, performance degrades with larger keys due to cubic complexity.

    - Java (Using JCA)

        - Time Complexity: $O(n^3)$.
        - Space Complexity: $O(n)$.
        - Execution Time: Less efficient than C++ especially as key size increases.

3. **Elliptic Curve Cryptography (ECC)**

   - C++ (Using OpenSSL)

     - Time Complexity: $O(n^2)$, where n is related to the field size.
     - Space Complexity: $O(n)$, depending on the curve and field size.
     - Execution Time: Very efficient, especially with optimized implementations for specific curves.

   - Java (Using JCA)

     - Time Complexity: $O(n^2)$.
     - Space Complexity: $O(n)$.
     - Execution Time: Improved in recent versions but still slower than C++.

4. **Learning with Errors (LWE) and Ring-LWE**

   - C++

     - Time Complexity: Polynomial in the dimension of the lattice, typically $O(n^{2.5})$ to $O(n^3)$.
     - Space Complexity: High, due to the need to manage large matrices or polynomials.
     - Execution Time: Fast with optimized libraries but varies greatly based on the specifics of the implementation and problem size.

   - Java

     - Time Complexity: Similar to C++ but generally exhibits worse constants due to less efficient low-level operations.
     - Space Complexity: Comparable to C++, but managed within the constraints of JVM.
     - Execution Time: Generally slower due to JVM overhead and less effective memory management for large data structures.

The performance obtained is shown in Table 2.4.

Table 2.4 Performance comparison for modern cryptography algorithms: AES, RSA, Elliptic Curve Cryptography, and (Ring) Learning-based Cryptography

Algorithm	Language	Time complexity	Space complexity	Execution time
AES	C++	$O(n)$	Constant	Fastest
AES	Java	$O(n)$	Constant	Fast
RSA	C++	$O(n^3)$	$O(n)$	Faster
RSA	Java	$O(n^3)$	$O(n)$	Slower
ECC	C++	$O(n^2)$	$O(n)$	Very efficient
ECC	Java	$O(n^2)$	$O(n)$	Efficient
LWE	C++	$O(n^{2.5})$–$O(n^3)$	High	Variable
LWE	Java	$O(n^{2.5})$–$O(n^3)$	High	Less efficient

Chapter 3
Quantum Algorithms

3.1 What are Quantum Algorithms?

Quantum computing is a discipline focused on the advancement of computer technology using the principles of quantum theory. *Quantum theory* is a branch of physics that elucidates the behavior of matter and energy at the atomic and subatomic scales.

Quantum computing, a rapidly evolving field at the intersection of computer science and quantum mechanics, encompasses several branches that focus on different aspects of technology and theory. Here are the main branches:

1. *Quantum Algorithms*. Focuses on the development of algorithms that run on quantum computers. These algorithms leverage quantum phenomena like superposition and entanglement to solve problems more efficiently than classical algorithms can.
2. *Quantum Hardware*. Involves the physical realization of quantum computers. This includes the development and refinement of qubits (quantum bits), which are the basic units of quantum information, and the various technological platforms used to implement them, such as superconducting circuits, trapped ions, topological qubits, and photonic systems.
3. *Quantum Error Correction*. Essential for building reliable quantum computers, this branch deals with overcoming and correcting errors that occur due to the fragile nature of quantum states. Quantum error correction schemes are crucial for achieving fault-tolerant quantum computation.
4. *Quantum Cryptography*. Exploits principles of quantum mechanics to perform cryptographic tasks that are provably secure against any adversary bounded by the laws of physics. The most well-known application is quantum key distribution, which allows two parties to generate a shared, secret random key.
5. *Quantum Communication*. Focuses on the transmission of quantum information between physically separated quantum computers or devices. Quantum communication schemes are vital for the development of quantum networks and long-distance secure communication.

M. I. Mihailescu et al., *Applied Quantum Computing and Cryptography*,
Studies in Big Data 167, https://doi.org/10.1007/978-3-031-77145-3_3

6. *Quantum Simulation.* Involves simulating quantum systems using quantum computers. This is particularly useful for understanding complex molecules and materials, which can have implications for chemistry and materials science.
7. *Quantum Machine Learning.* Combines quantum computing with machine learning techniques. Quantum machine learning explores how quantum algorithms can improve aspects of data processing and pattern recognition.
8. *Quantum Metrology and Sensing.* Uses quantum properties to improve measurement precision beyond what is possible with classical physics. This branch has applications in various fields, including gravitational wave detection and magnetic field sensing.

Each of these branches plays a crucial role in the advancement of quantum computing, contributing to both theoretical and practical breakthroughs.

3.1.1 Background

Nobel laureate Richard Feynman introduced the concept of a quantum computer in 1981. He stated that a classical computer could not reliably and effectively replicate quantum mechanical systems, but that a new type of machine—a computer "built of quantum mechanical elements that obey quantum mechanical laws"—might one day be able to do so. Because the amount of information required to fully represent a quantum system grows exponentially, classical computers are fundamentally incapable of simulating such a system with sub-exponential time and spatial complexity. Quantum computers, on the other hand, make use of the unique, non-classical properties of the quantum systems from which they are created, allowing them to process exponentially large quantities of data in only polynomial time. It goes without saying that this kind of computer capability could be employed to solve a variety of problems unrelated to quantum mechanics. Similarly to how classical computation quickly expanded beyond its initial focus on simulating Newtonian mechanics, quantum algorithm research has had a significant impact on a wide range of fields, including information theory, cryptography, language theory, and mathematics.

Peter Shor's 1997 publication of a quantum algorithm for completing prime factorization of integers in almost polynomial time is probably the most well-known advancement in quantum computation. Not only does Shor's algorithm outperform the fastest classical methods exponentially, but many public-key cryptography algorithms, including the widely used RSA algorithm, are based on the fact that there is currently no effective classical solution for factoring integers into prime numbers. Shor demonstrated how the construction of a full-scale quantum computer may result in a genuinely massive increase in computing speed, while also highlighting the possible repercussions of such an increase in computing power on the cybersecurity sector. For both of these reasons, Shor's discovery sparked a great deal of interest, which is still evident in the development of quantum algorithms and computers.

Aside from Shor's algorithm, there have been several fascinating and noteworthy algorithms developed for quantum computers. In this tutorial, two of these algorithms will be thoroughly presented in order to help students understand quantum computing theory and quantum algorithm construction. Many other quantum algorithms, whether employed to solve linear equation systems or quickly compute discrete logarithms, use similar methodologies to get their results, so these two algorithms serve as good examples for our present understanding of quantum processing. For example, Shor believes that one of these algorithms served as the basis for his own. The lecture will start with a look at Lov Grover's quantum database search technique. Grover's algorithm outperforms the best classical algorithms by searching for a specific entry in an unordered database while employing amplitude amplification, an important quantum algorithm design method. Grover's algorithm is, in fact, the most effective quantum algorithm for conducting such a search. Of course, identifying a single element in an unordered collection can be used to a variety of computer science problems, including the SAT problem of boolean satisfiability. Finding an element, the set of truth values, that satisfies a specific condition, the set of clauses, in an unordered search space, the set of all 2n possible truth assignments of n boolean variables, is the same as determining whether a given set of clauses is satisfied. Even if this algorithm does not much exceed its traditional counterparts in terms of efficiency, the speedup is still noticeable with large input.

The second technique in this course will examine Daniel Simon's approach to discovering the exclusive-or (XOR) mask across which a particular black-box function is invariant. Simon's approach has an optimal runtime and was the first quantum algorithm discovered to perform exponentially faster than any matching classical algorithm. Although Simon's problem looks to be fairly abstract, the hidden Abelian subgroup problem, which includes factoring integers into primes and computing discrete logarithms, is an important problem with several applications in computer science.

3.1.2 Challenges and Warnings

Why aren't quantum computers more widely used if they're so much faster than conventional machines? Even while engineering has evolved substantially, developing a quantum computer remains a daunting undertaking. The circuit was hardwired to only factor the number 15, not any other input, as the approach is intended to do. For example, the maximum number factored by a quantum computer using Shor's technique is 15. To put this in context, Shor's approach would need to be applied to a 1024-bit integer in order to crack 1024-bit RSA keys, which are currently recommended for corporate use. This would necessitate a quantum computer orders of magnitude more powerful than that needed to divide 15 by 5 and 3.

The specifics of the various physical quantum computer implementations that engineers are now working with are outside the scope of this course. The most extensively utilized technique at the present, a "ion trap quantum computer," uses lasers to modify the spin states of supercooled ions trapped in an electromagnetic field. This method has produced some of the largest quantum machines, with registers that can hold up to 8 quantum bits (qubits). All of the concepts considered thus far share a common challenge: creating a machine that is sensitive to user or engineer control while remaining unaffected by its surroundings. The decoherence of a quantum computer implementation, caused by the inherent coupling of man-made quantum systems to the unpredictable natural world that we use to make them, is used to evaluate the system's quality. For example, in the ion trap implementation, any adjacent electric and magnetic forces, even those required to power the device, may generate undesirable interference by causing the ions to arbitrarily change spin states before collapsing into a classical system. While decreasing these sources of decoherence is the primary goal of current research, it cannot be completely avoided. As a result, while decoherence can currently be reduced to a level that allows for reasonably accurate computation on the scale of 8 qubits, it increases with the number of qubits, to the point where properly functioning quantum computers with more than 8 qubits have yet to be realized.

Nonetheless, numerous academic institutions and business research facilities around the world are working to develop the first scalable quantum computer because quantum computation—despite the fact that classical machines have failed to match its performance promises—remains a strong contender for the next computational paradigm. Even if general-purpose quantum computers appear to be a long way off, small-scale quantum computers are expected to be employed by classical computers to speed up particular computations. Examples include the calculations required to mimic complicated biological systems or to determine the content of a raster image for improved online search. To be prepared for the quantum computational tools that will most likely be available in the not-too-distant future, you must first learn the fundamentals of quantum computation.

3.1.3 Limitations of Classical Computing

First, let's define classical computing. Essentially, it requires binary processing, which stores data in bits. A bit is always in one of two distinct physical states, represented by a single binary value, usually expressed as 0 or 1. Bits are physically stored in memory via transistors connected in a circuit. The device's intrinsic capacitance allows it to store electrical charge. Each bit's charge defines its state, which, in turn, determines its value. Transistors are frequently made using semiconductor materials such as silicon.

The MOSFET (see Fig. 3.1), also known as the metal–oxide–semiconductor field-effect transistor, is an insulated-gate field-effect transistor (IGFET) that is created by deliberately oxidizing a semiconductor material, usually silicon. The electrical conductivity of the device is determined by the voltage provided to the covered gate. This capacity to modify conductivity based on the applied voltage can be utilized for amplifying or switching electronic signals.

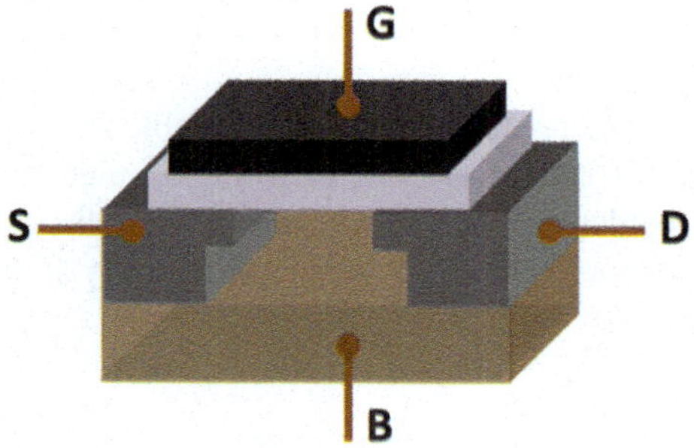

Fig. 3.1 MOSFET. We have the gate *(G)*, body *(B)*, the source *(S)*, and *drain (D)* terminals. The gate is separated from the body by an insulating layer (pink) (*Source of image*: MOSFET, link: https://en.wikipedia.org/wiki/List_of_MOSFET_applications. Last access: May 25th, 2024)

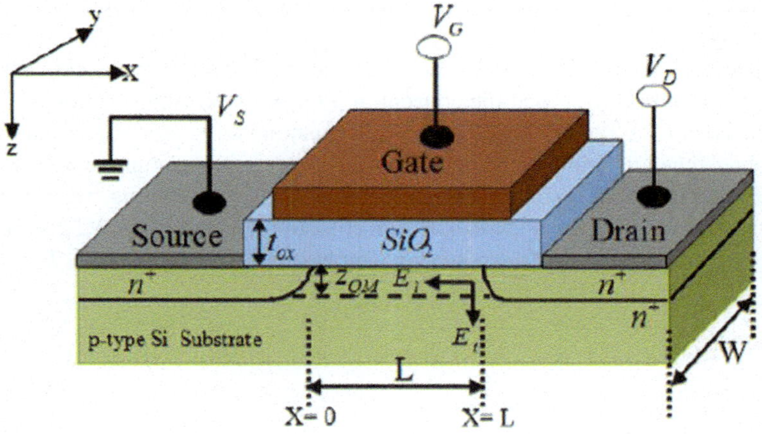

Fig. 3.2 The main structure for n-channel MOSFET that has the electrons removed from the Si/SiO_2 interface because of the the quantum-confinement effect (*Source of image*: Saad, Ismail & Tan, Michael & Ahmadi, Mohammad & Ismail, Razali & Arora, Vijay. (2010). The dependence of saturation velocity on temperature, inversion charge, and electric field in a nanoscale MOSFET. Int. J. Nanoelectronics and Materials. 3. https://www.researchgate.net/publication/229021006_The_dependence_of_saturation_velocity_on_temperature_inversion_charge_and_electric_field_in_a_nanoscale_MOSFET. Last access: May 25th, 2024)

Figure 3.2 shows a nanoscale channel with a few nanometers of oxide thickness. The quantum-confinement effect separates electrons from the interface, which is an important element to consider. Because it does not offer acceleration, the gate electric field, E_t, produces no heat in electrons. Instead, it acts as a confined electric field, transforming an electron into a quantum entity with wave characteristics and discrete energy states.

A microprocessor is an integrated circuit that performs the function of executing instructions in traditional computing. Typically, it comprises of billions of transistors grouped into thousands of discrete digital circuits, each performing a unique logical operation.

CPUs have become more inexpensive and capable over the last few decades, as predicted by Moore's Law. However, there is a limited limit to the number of transistors that may be densely integrated into a digital circuit before reaching the limits imposed by fundamental physical laws, particularly at the subatomic level.

3.1.4 The Quantum Computing Birth

Quantum computing is based on the concepts of quantum mechanics, a scientific theory developed in the early twentieth century before the widespread use of classical computer. For a long time, quantum physics and computer science were studied separately.

The convergence of these areas did not occur until the late twentieth century. In 1980, Paul Benioff proposed the concept of a Quantum Turing Machine (QTM) as an extension of a Turing Machine (TM). A quantum simulator is a computational device designed to simulate the behavior and effects of a quantum computer.

For similarities of quantum computing with classic computing, we can see the Turing Machine comparison with quantum computing in Table 3.1. We start from the representation of a Turing Machine as a 7-tuple structure: $M = (Q, \Sigma, \Gamma, \delta, B, F)$.

Table 3.1 Machine turing model versus quantum computer model

Turing machine	Quantum computer
Q: Represents the finite set of states for the finite control	The set of states Q is replaced by the Hilbert Space
Σ: The finite set of input symbols	The input and output symbols Σ are usually taken as a discrete set, similar with the classical
Γ: The finite set of tape symbols, Σ and Γ	The tape alphabet Γ is likewise replaced with a Hilbert space (usually a different Hilbert space than the set of states)
δ: The transition function	The transition function $\delta : \Sigma \bigotimes \Gamma \rightarrow \Sigma \times Q \bigotimes \Gamma \times L, R$
$q_0 \in Q$ represents the start state	The initial state $q_0 \in Q$ could be a mixed state or a pure state
$B \in \Sigma$ is the blank symbol, $B \notin \Sigma$	The blank symbol $b \in \Gamma$ represents an element from the Hilbert space
$F \subseteq Q$ represents the set of the final or accepting states	The set F of final or accepting states is a subspace of the Hilbert Space Q

Simulating quantum dynamics with classical computers was challenging, despite the numerous breakthroughs. Subsequently, during the 1980s, Richard Feynman and Yuri Manin put forth separate proposals indicating that hardware utilizing quantum phenomena could potentially offer greater efficiency for these simulations.

A qubit, also known as a quantum bit, is the fundamental unit of information in quantum computing, much like the classical bit in regular computing. Qubits, unlike bits, can exist in both of their fundamental states at the same time. A qubit is a binary quantum-mechanical system that exists in two different states.

3.1.5 Quantum Mechanics Concepts

The development of quantum mechanics spans a period of more than two centuries and involves the contributions of numerous experts. Classical mechanics effectively explained the workings of the natural world by employing the principles of motion and gravity. Nevertheless, the scientific investigation into the characteristics of light was insufficient in providing an explanation based on the prevailing views of that era.

3.1.5.1 Initial Quantum Theory Period

Isaac Newton developed the basic concepts of the corpuscular theory of light. However, contradictions discovered in studies encouraged the development of the wave theory of light. The famous interference experiment performed by Thomas Young in 1803 provides a clear and complete description of the phenomena.

In 1859, Gustav Kirchhoff presented a description of the black-body radiation problem, which posed a challenge to the wave theory of light. In 1900, Max Planck proposed the concept of discrete "quanta" to explain the emission and absorption of energy in black-body radiation. This is shown in Eq. 3.1.

$$E = h \cdot v \tag{3.1}$$

where

- E is the energy of a photon;
- v is the frequency of the photon;
- h represents the Planck's constant = 6.626×10^{-34} J-s.

However, Planck did not take his quantum idea very seriously. In 1905, Albert Einstein used Planck's quantum hypobook to explain the photoelectric effect, which is the emission of electrons when electromagnetic radiation interacts with a substance.

In 1924, Louis-Victor de Broglie proposed the hypobook that all matter has wave-like characteristics. The experimental observation of electron diffraction quickly provided evidence for this. Subsequently, an examination was conducted on bigger particles such as neutrons and protons are shown in Eq. 3.2.

$$\lambda = \frac{h}{p} \tag{3.2}$$

where

- λ is de Broglie wavelength associated with a massive particle.
- p represents the momentum of particle.
- h represents the Planck's constant.

The scientific advancements made by researchers between 1900 and 1925 are commonly known as the "old quantum theory". Essentially, it is a compilation of heuristic adjustments to classical physics, rather than a comprehensive theory of quantum mechanics.

3.1.5.2 Modern Period of Quantum Theory

Following 1925, a comprehensive set of mathematical rules was developed that could accurately represent and combine several approaches to quantum theory. Erwin Schrödinger, Werner Heisenberg, John von Neuman, Paul Dirac, and other renowned scientists pioneered the development of the "new quantum theory".

In 1926, Erwin Schrödinger developed a linear partial differential equation to explain the wave function of a quantum-mechanical system, as illustrated in Eq. 3.3. This was a significant step forward in quantum mechanics.

$$i\hbar\frac{\partial}{\partial t}|\Psi t\rangle = \hat{H}|\Psi t\rangle \tag{3.3}$$

where

- i represents the imaginary unit.
- $\hbar$ represents the reduced Planck's constant.
- t represents duration.
- $|\Psi t\rangle$ represents the quantum state vector.
- $\hat{H}$ represents the Hamiltonian operator.

Another key discovery came in 1927, when Werner Heisenberg proposed his uncertainty principle. This concept limits the precision with which specific pairs of physical properties of a particle may be determined based on its initial conditions, as stated in Eq. 3.4.

$$\Delta x \Delta p \geq \frac{h}{4\pi} \tag{3.4}$$

where

- Δx represents the uncertainty in position.
- Δp represents the uncertainty of momentum.
- h represents the Planck's constant.

These and other scientific research have helped to shape the concept of wave-particle duality, which argues that any quantum occurrence can be described as a particle or a wave. Despite providing answers for a vast range of observed events, it is a matter of constant bewilderment in the field of modern physics.

From approximately 1930 onwards, the concepts derived from the emerging quantum theory were utilized to develop the Quantum Field Theory () in the field of electromagnetic. The theoretical framework combines classical field theory, special relativity, and quantum mechanics. It has led in the development of numerous formulations, including the Standard Model.

3.1.6 The Principles of Quantum Mechanics

An in-depth explanation of a quantum-mechanical system is a complex topic that exceeds the limits of this lesson. Nevertheless, the coexistence of wave and particle properties in a quantum item gives rise to other intriguing physical phenomena. These ideas serve as the fundamental basis of quantum information science.

3.1.6.1 Quantum Superposition

Quantum superposition (Fig. 3.3) is a fundamental premise of quantum physics. It states that a particle can be in a superposition of several quantum states, or eigenstates. This is in contrast to classical mechanics, where concepts like as location and momentum are always fully defined.

In mathematics, multiple quantum states can be combined or joined to form a new, valid quantum state. In contrast, we can express a quantum state as a mixture of two or more independent states. It refers to the properties of solutions to .

Surprisingly, when the wave function interacts with the outside world, it goes through a process known as wave function collapse, which reduces it to a single eigenstate. This provides a relationship between the wave function and classical observables like position and momentum. Furthermore, this provides the foundation for quantifying phenomena in the field of quantum mechanics.

3.1.6.2 Quantum Entanglement

Quantum entanglement (Fig. 3.4) occurs when a collection of particles is created in such a way that the quantum state of each individual particle cannot be characterized

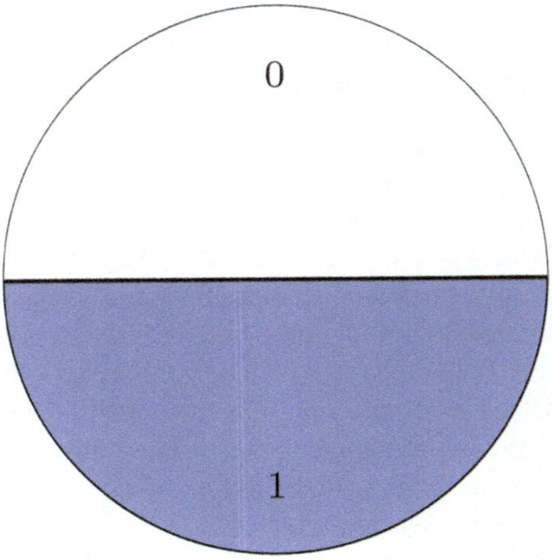

Fig. 3.3 The qubit that is found in a superposition state with the probability equal of collapsing into either of the basis states, 0 and 1

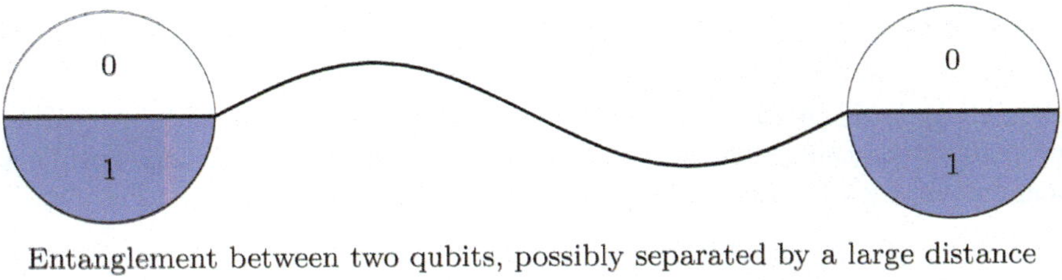

Fig. 3.4 Illustration of quantum entanglement

independently of the states of the other particles, even when they are physically separated by a significant distance.

Consequently, when entangled particles are measured for physical attributes like location and momentum, there can be instances where the measurements are completely coupled. It is crucial because the entangled particles' mutual information can be utilized.

Throughout history, the issue of entanglement has sparked worry. The Einstein-Podolsky-Rosen dilemma demonstrates the limitations of the wave function's quantum-mechanical representation of physical reality. Nevertheless, empirical investigations have revealed evidence of entanglement.

3.1.7 The Main Element of Quantum Computing

Although quantum mechanics has made tremendous progress in understanding the majority of natural events, the idea of gravity remains elusive and is best understood within the context of general relativity. However, one notable outcome of this attempt

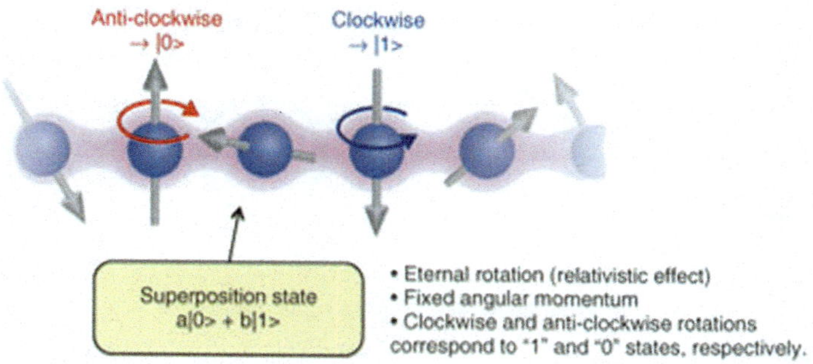

Fig. 3.5 The diagram of electron spin (*Source of image*: https://www.ntt-review.jp/archive/ntttechnical.php?contents=ntr201209fa1.html. Last access: May 29th, 2024)

has been the creation of quantum information science, which combines quantum mechanics and information technology.

3.1.7.1 Qubit

A qubit is the fundamental unit of quantum information. Quantum information is the data that describes the specific condition or state of a quantum system. Furthermore, a qubit is a quantum-mechanical particle that can exist in two different states. This system has the potential to become a quantum particle, such as an electron or a photon. A photon's quantum property is polarization, which can exist in two different quantum states: vertical and horizontal. Similarly, an electron's spin (see Fig. 3.5) can exist in two separate quantum states: spin-up or spin-down.

A qubit can exist in a quantum superposition, which implies it can be in any of its two unique quantum states at the same time. Each quantum state has an associated probability amplitude. When a measurement is performed, the qubit's state collapses into one of two potential quantum states.

Quantum interference changes the state of a qubit, which influences the likelihood of a given result during measurement. Although the features of an individual qubit are intriguing, the full potential of quantum computing resides in the usage of several qubits, establishing a multi-qubit system (see Fig. 3.6).

Quantum entanglement allows qubits that are far apart to interact with each other instantly. This allows numerous qubits to work together and create a global system. It works as a computational amplifier for qubits.

Thus, the quantum computer can execute 2^n computations, where n is the number of qubits employed. The combined quantum properties of superposition and entanglement result in a huge improvement in computational capability.

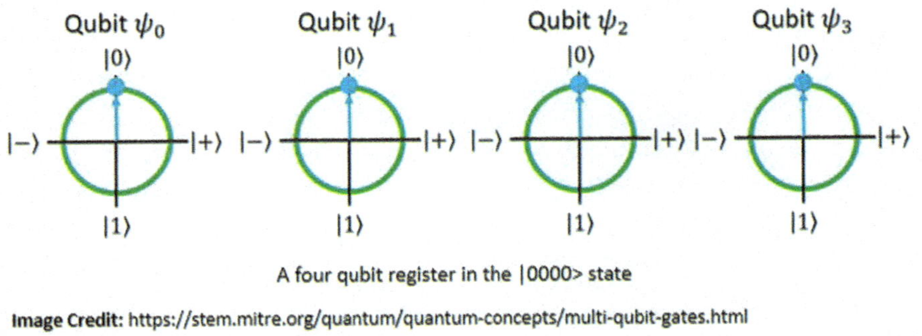

Fig. 3.6 A four qubit register in the $|0000\rangle$ (*Source of image*: A Multi-Qubit Gate: the SWAP Gate https://stem.mitre.org/quantum/quantum-concepts/multi-qubit-gates.html. Last access: May 29th, 2024)

3.1.7.2 Representation of the Qubit

Because a qubit can exist in a superposition of potential quantum states, the most accurate approach to represent its state is using a two-dimensional column vector. The quantum state vector has a unit norm, which means that the squared magnitudes of its elements must sum up to one.

For example, $\begin{bmatrix} \alpha \\ \beta \end{bmatrix}$ represents a valid qubit state as shown in Eq. (3.5)

$$|\alpha|^2 + |\beta|^2 = 1 \tag{3.5}$$

In Eq. (3.5), α and β represent a real or complex numbers.

The quantum state vectors $\begin{bmatrix} 1 \\ 0 \end{bmatrix}$ and $\begin{bmatrix} 0 \\ 1 \end{bmatrix}$ represents the basis vectors and the functionality that corresponds to the two states of a classical bit. Further, we will present a form for a quantum state vector as being represented as the sum of each basis vectors that has been observed in our discussion so far (see Eq. 3.6).

$$\begin{bmatrix} x \\ y \end{bmatrix} = x \begin{bmatrix} 1 \\ 0 \end{bmatrix} + y \begin{bmatrix} 0 \\ 1 \end{bmatrix} \tag{3.6}$$

Next, we will measure the quantum state vector as in Eq. 3.7.

$$\begin{bmatrix} \alpha \\ \beta \end{bmatrix} \tag{3.7}$$

In Eq. 3.5 the probability of getting the outcome 0 is α^2, and the probability of obtaining the outcome 1 is β^2. As a result, the qubit's state changes to either [1, 0] or [0, 1].

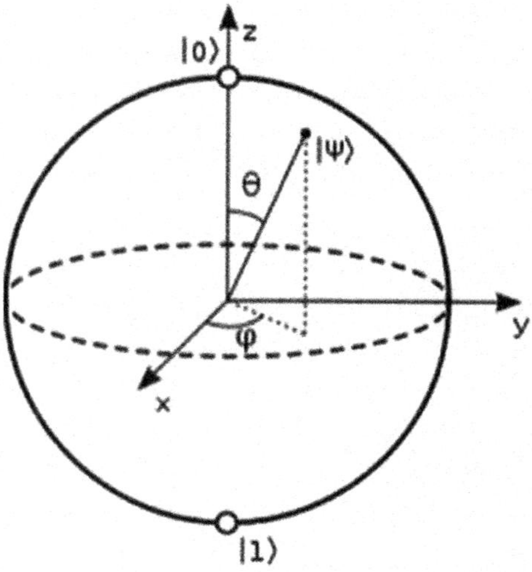

Fig. 3.7 The Bloch sphere

3.1.7.3 A Different Representation

Although row and column vector notation is widely used in linear algebra, it is unsuitable for quantum computing. This is especially accurate when dealing with several qubits. This leads to uncertainty in the vector space and complicates the representation of even simple vectors.

Thus, Dirac notation is frequently used in quantum mechanics to denote quantum states. Paul Dirac invented the bra-ket notation. The notation uses angle brackets $\leq$ and vertical bars | to create "bras" and "kets". The single-qubit basis states can be described using Dirac's notation, as shown in Eq. 3.8.

$$\begin{bmatrix} 1 \\ 0 \end{bmatrix} = |0\rangle, \begin{bmatrix} 0 \\ 1 \end{bmatrix} = |1\rangle \tag{3.8}$$

If ϕ is a row vector, it can be expressed as $\langle\phi|$. Similarly, if ψ is a column vector, we can define it as a ket-vector indicated by $|\psi\rangle$. Assuming ϕ and ψ are quantum state vectors, $\langle\phi||\psi\rangle$ represents their inner product.

Felix Bloch, a scientist, devised the Bloch sphere (see Fig. 3.7), a geometric portrayal of the space that reflects the pure states of a two-level quantum mechanical system. The object in question is simply a unit 2-sphere with opposite points representing the orthogonal state vectors.

3.1.7.4 Quantum Operations and Linear Algebra

Linear algebra serves as the primary mathematical framework for representing and communicating concepts in quantum computing. The term is commonly employed to characterize qubit states and quantum operations, as well as to anticipate the behavior of a quantum computer that reacts to a set of instructions.

We have seen how to represent the state of a single qubit using vectors. The quantum state of two qubits can be written as the tensor product of their distinct state vectors, as demonstrated in Eq. (3.9).

$$\begin{bmatrix} a \\ b \end{bmatrix} \otimes \begin{bmatrix} c \\ d \end{bmatrix} = \begin{bmatrix} a = \begin{bmatrix} c \\ d \end{bmatrix} \\ b = \begin{bmatrix} c \\ d \end{bmatrix} \end{bmatrix} = \begin{bmatrix} ac \\ ad \\ bc \\ bd \end{bmatrix} \tag{3.9}$$

Quantum operations can be represented using a matrix. When a quantum operation is applied to a qubit, the resulting state of the qubit can be obtained by multiplying the two matrices as shown in Eq. (3.10).

$$H = \frac{1}{\sqrt{2}} \begin{bmatrix} 1 & 1 \\ 1 & -1 \end{bmatrix} H \begin{bmatrix} a \\ b \end{bmatrix} = \frac{1}{\sqrt{2}} = \begin{bmatrix} 1 & 1 \\ 1 & -1 \end{bmatrix} \begin{bmatrix} a \\ b \end{bmatrix} = \frac{1}{\sqrt{2}} \begin{bmatrix} 1 \\ 1 \end{bmatrix} \tag{3.10}$$

3.1.7.5 Quantum Gates and Circuits

There are various methodologies for constructing quantum computers using the fundamental component of qubits. Quantum circuits are a model of computation that involve a succession of quantum gates, measurements, qubit initialization, and maybe other operations.

A quantum logic gate is a fundamental quantum circuit that operates on a limited number of qubits. It is essentially the same as classical logic gates seen in traditional digital circuits. Contrary to conventional logic gates, quantum logic gates possess the property of reversibility, see Figs. 3.8 and 3.9, and Tables 3.2 and 3.3.

Quantum gates are represented by unitary operators, which can be described as unitary matrices in terms of a certain basis, usually the computational basis. The

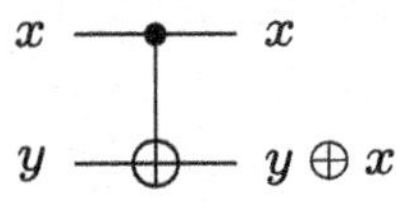

Fig. 3.8 Controlled NOT (CNOT) gate or Feynman Gate

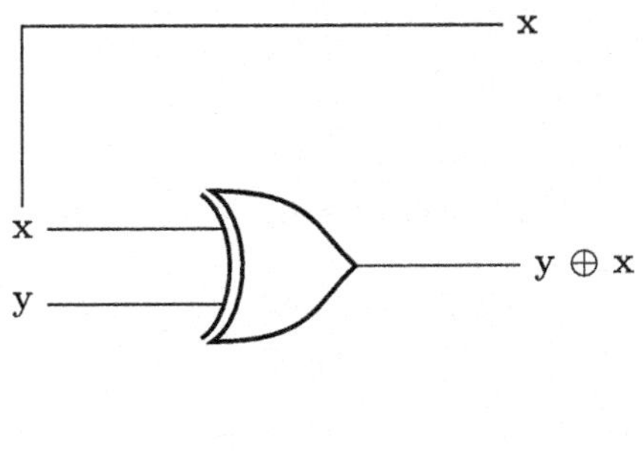

Fig. 3.9 XOR Gate

Table 3.2 The inputs and outputs for Controlled NOT (CNOT) Gate

Input		Output	
x	y	x	y + x
$\|0\rangle$	$\|0\rangle$	$\|0\rangle$	$\|0\rangle$
$\|0\rangle$	$\|1\rangle$	$\|0\rangle$	$\|1\rangle$
$\|1\rangle$	$\|0\rangle$	$\|1\rangle$	$\|1\rangle$
$\|1\rangle$	$\|1\rangle$	$\|1\rangle$	$\|1\rangle$

Table 3.3 The inputs and outputs for XOR Gate

Input		Output	
x	y	x	y + x
0	0	0	0
0	1	0	1
1	0	1	1
1	1	1	0

current representation of quantum gates is based on the notation first proposed by Richard Feynman in 1986.

An example of a fundamental quantum logic gate is the Identity gate, which may be defined for a single qubit as indicated in Eq. (3.11).

The I from Eq. (3.11) represents a basis-independent operator that does not alter the quantum state. Various researchers (see Table 3.4) have defined several other quantum gates.

$$I = \begin{bmatrix} 1 & 0 \\ 0 & 1 \end{bmatrix} \tag{3.11}$$

DiVincenzo's criteria refers to the essential set of activities that a quantum circuit must be capable of executing on the qubits in order to provide quantum computation. In 2000, David P. DiVincenzo put out a proposal consisting of seven characteristics that an experiment must meet in order to successfully carry out quantum algorithms.

The description of quantum circuit components is illustrated using a modified form of Roger Penrose's graphical notation, which is a visual representation of tensors established in 1971. Figure 3.10 depicts a quantum circuit used for qubit teleportation.

The flow of information from left to right in a quantum circuit represents the progression of time, resulting in the gates being ordered chronologically. The quantum logic gate represented by 'H' is a Hadamard operation that acts on a register with a single qubit. The other function performed in this circuit is measurement, which is represented by the meter symbol.

Table 3.4 The inputs and outputs for XOR Gate

Operator	Gate(s)	Matrix
Pauli-X (X)	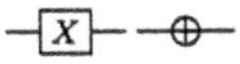	$\begin{bmatrix} 0 & 1 \\ 1 & 0 \end{bmatrix}$
Pauli-Y (Y)	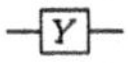	$\begin{bmatrix} 0 & -i \\ i & 0 \end{bmatrix}$
Pauli-Z (Z)		$\begin{bmatrix} 1 & 0 \\ 0 & -1 \end{bmatrix}$
Hadamard (H)	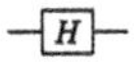	$\frac{1}{\sqrt{2}}\begin{bmatrix} 1 & 1 \\ 1 & -1 \end{bmatrix}$
Phase (S, P)	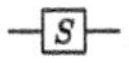	$\begin{bmatrix} 1 & 0 \\ 0 & i \end{bmatrix}$
$\pi/8$ (T)	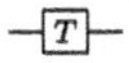	$\begin{bmatrix} 1 & 0 \\ 0 & e^{i\pi/4} \end{bmatrix}$
Controlled Not (CNOT, CX)		$\begin{bmatrix} 1 & 0 & 0 & 0 \\ 0 & 1 & 0 & 0 \\ 0 & 0 & 0 & 1 \\ 0 & 0 & 1 & 0 \end{bmatrix}$
Controlled Z (CZ)	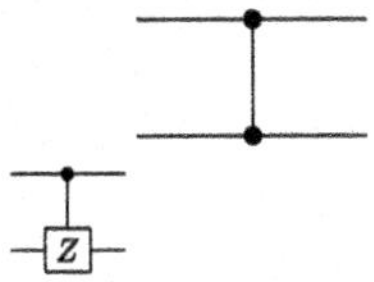	$\begin{bmatrix} 1 & 0 & 0 & 0 \\ 0 & 1 & 0 & 0 \\ 0 & 0 & 1 & 0 \\ 0 & 0 & 0 & -1 \end{bmatrix}$
SWAP	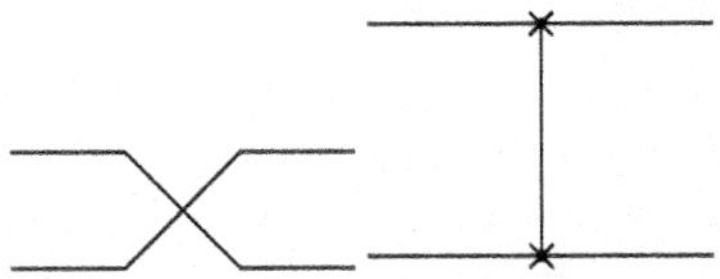	$\begin{bmatrix} 1 & 0 & 0 & 0 \\ 0 & 0 & 1 & 0 \\ 0 & 1 & 0 & 0 \\ 0 & 0 & 0 & 1 \end{bmatrix}$
Toffoli (CCNOT, CCX, TOFF)	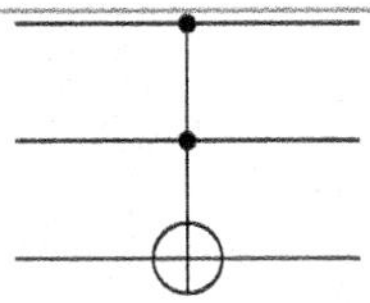	$\begin{bmatrix} 1 & 0 & 0 & 0 & 0 & 0 & 0 & 0 \\ 0 & 1 & 0 & 0 & 0 & 0 & 0 & 0 \\ 0 & 0 & 1 & 0 & 0 & 0 & 0 & 0 \\ 0 & 0 & 0 & 1 & 0 & 0 & 0 & 0 \\ 0 & 0 & 0 & 0 & 1 & 0 & 0 & 0 \\ 0 & 0 & 0 & 0 & 0 & 1 & 0 & 0 \\ 0 & 0 & 0 & 0 & 0 & 0 & 0 & 1 \\ 0 & 0 & 0 & 0 & 0 & 0 & 1 & 0 \end{bmatrix}$

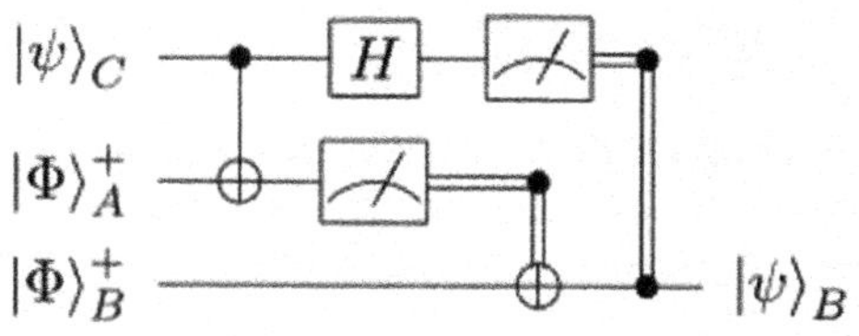

Fig. 3.10 Quantum circuits that executes qubit teleportation (*Source of image*: Quantum circuit https://en.wikipedia.org/wiki/Quantum_circuit. Last access: May 27th, 2024)

3.2 Quantum Fourier Transform

The Quantum Fourier Transform (QFT) [1] is one of the most important and powerful operations in quantum computing. It is the quantum version of the classical discrete Fourier transform (DFT) and has an important role in different quantum algorithms, including Shor's algorithm (see Sect. 3.5) for integer factorization and quantum phase estimation.

The classical Fourier transform is a mathematical technique that transforms a sequence of values into components of different frequencies. It is widely used in signal processing, image analysis, and many other fields. The Quantum Fourier Transform achieves a similar transformation on quantum states, leading to operations that are exponentially faster than their classical versions for specific problems.

The QFT is applied on a quantum state $|\psi\rangle$ which is a superposition of 2^n basis states in an n-qubit system. Mathematically, the QFT is defined as follows: given an n-qubit state

$$|\psi\rangle = \sum_{k=0}^{2^n-1} x_k|k\rangle, \tag{3.12}$$

the QFT transforms this state into:

$$QFT(|\psi\rangle) = \frac{1}{\sqrt{2^n}} \sum_{j=0}^{2^n-1} \Big(\sum_{k=0}^{2^n-1} x_k e^{2\pi ijk/2^n}\Big)|j\rangle, \tag{3.13}$$

In the matrix form, the QFT is represented by a unitary matrix F with elements

$$F_{jk} = \frac{1}{\sqrt{2^n}} e^{2\pi ijk/2^n} \tag{3.14}$$

3.2.1 *Quantum Fourier Transform Steps*

The QFT algorithm can be implemented using a sequence of Hadamard gates and controlled phase shift gates. It has four main steps on which we describe below.

- **Initialization**. Begin with an n-qubit quantum register in the state $|\psi\rangle = \sum_{k=0}^{2^n-1} x_k|k\rangle$.
- **Hadamard gates**. The Hadamard gate H is applied to each qubit, creating a superposition of all basis states. For the j-th qubit, this operation is:

$$H|k\rangle = \frac{1}{\sqrt{2^n}}\left(|0\rangle e^{2\pi i\cdot 0\cdot k/2^1}|1\rangle\right) \tag{3.15}$$

- **Controlled phase shift gates**. Controlled phase shift gates R_m introduce relative phases between the basis states. The controlled-R_m gate is applied on two qubits, where m represents the level of control. For qubits q and p, with $q < p$, the phase shift is:

$$R_m|k\rangle = e^{2\pi i/2^m}|k\rangle \tag{3.16}$$

  This step introduces the necessary phase relationships between the qubits.
- **Reverse qubit order**. After applying the Hadamard and controlled phase shift gates, the order of the qubits is reversed to complete the QFT. This step ensures the correct mapping from the original basis states to the transformed states.

The QFT can be visualized using a quantum circuit. Below, we present a simple example of the circuit for a 3-qubit system:

1. **Hadamard on q_0**: Apply H to the first qubit q_0.
2. **Controlled Phase Shifts on q_0**: Apply R_2 (controlled-phase) between q_0 and q_1, then R_3 between q_0 and q_2.
3. **Hadamard on q_1**: Apply H to the next qubit q_1.
4. **Controlled Phase Shift on q_1**: Apply R_2 (controlled-phase) between q_1 and q_2.
5. **Hadamard on q_2**: Apply H to the last qubit q_2.
6. **Reverse Order**: Swap the positions of the qubits to reverse their order.

3.2.2 Implementation

The implementation for the Quantum Fourier Transform is given in Listing and follows the steps presented in Sect. 3.2.1, while the result is shown in Fig. 3.11.

```
from qiskit_aer import AerSimulator
from qiskit import QuantumCircuit, transpile, assemble
from qiskit.visualization import plot_histogram, plot_bloch_multivector
import numpy as np
import matplotlib.pyplot as plt

def qft(qc, n):
"""Apply the Quantum Fourier Transform to the first n qubits in the circuit qc."""
# Apply the QFT to the first n qubits
for qubit in range(n):
# Apply the Hadamard gate to the qubit
qc.h(qubit)
# Apply the controlled phase shift gates
for k in range(2, n - qubit + 1):
qc.cp(np.pi / 2**(k - 1), qubit, qubit + k - 1)
# Reverse the order of the qubits
```

```
for qubit in range(n // 2):
qc.swap(qubit, n - qubit - 1)
return qc

# Function to create a QFT circuit
def create_qft_circuit(n):
qc = QuantumCircuit(n)
qft(qc, n)
qc.measure_all()
return qc

# Number of qubits
n_qubits = 3

# Create the QFT circuit
qft_circuit = create_qft_circuit(n_qubits)

# Visualize the circuit
print(qft_circuit)

backend = AerSimulator()
job = backend.run(qft_circuit, shots=1)
results = job.result()

# Get the measurement outcomes
counts = results.get_counts(qft_circuit)

# Show the results
print("Measurement results:", counts)
```

Listing 3.1 Implementation of the QFT

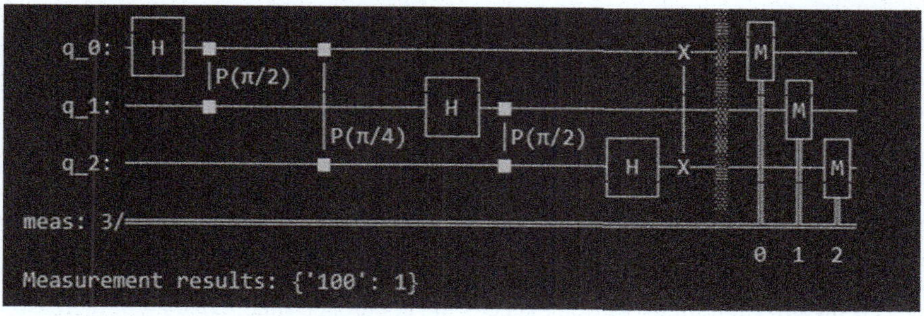

Fig. 3.11 QFT Python output

3.2.3 *Performance Analysis*

The QFT is an important quantum operation with applications in several key quantum algorithms, such as Shor's algorithm. Its performance can be evaluated based on three primary metrics: time complexity, space complexity, and execution time. Below we detail the analysis of each metric.

Time complexity. Time complexity shows the algorithm's efficiency, indicating how the computational resources required scale with the size of the input. For, QFT on n qubits, the time complexity operation is given by the Hadamard gates and controlled phase shift gates, resulting in a total complexity of $O(n^2)$, as follows:

- **Hadamard Gates**: Each qubit in the QFT undergoes a Hadamard transformation. For an n-qubit system, there are n Hadamard gates.
- **Controlled Phase Shift Gates**: For each qubit i controlled phase shifts gates R_k are applied to all subsequent qubits j, with $j > i$. This means that qubit i interacts with qubits from range $i + 1, \ldots, n - 1$. The total number of controlled phase shift gates is given by:

$$\sum_{i=0}^{n-1}(n - i - 1) = \frac{n(n-1)}{2} \tag{3.17}$$

The time complexity $O(n^2)$ of QFT is exponentially more efficient that the classical Fourier Transform, whose complexity is $O(n2^n)$ for a number represented on n bits.

Execution time is an empirical measure that depends on the specifics of the quantum hardware and the efficiency of the implementation. Several factors affect the execution time:

- **Gate Times**
  - **Hadamard Gates**: Usually very fast, on the order of nanoseconds.
  - **Controlled Phase Shift Gates**: Slightly more complex, as they are multi-qubit gates, but still on the order of nanoseconds to microseconds.
- **Number of gates**. As established, the total number of gates is $O(n^2)$, which influences the execution time linearly with the number of gates.
- **Quantum hardware limitations**
  - **Coherence Time**: The time for which a quantum state can maintain its integrity. If the QFT circuit execution exceeds the coherence time, errors will increase.
  - **Gate Fidelity**: Real quantum gates have imperfections. Higher fidelity gates reduce the error rate but might take longer to execute.
  - **Simulator**: The execution time is dominated by the classical computation resources available.
  - **Real Hardware**: Execution time is influenced by gate speeds and the need for error correction.

Space complexity. Space complexity measures the amount of memory required for the algorithm to run, which in the case of quantum computing translates to the number of qubits and classical bits used, as follows:

- **Quantum space**: The QFT is applied on n qubits to represent an n-bit input state. Therefore, the space complexity in terms of quantum bits is $O(n)$.
- **Classical space**: If measurement is required, n classical bits are needed to store the result of the measurement. Thus, the classical space complexity is also $O(n)$.

Overall, the space complexity of the QFT is $O(n)$ qubits and $O(n)$ classical bits, making it very efficient in terms of memory requirements.

3.3 Quantum Phase Estimation

Quantum Phase Estimation (QPE) [2] is a fundamental algorithm in quantum computing, being used in different quantum algorithms such as Shor's factoring algorithm, quantum simulations, and solving eigenvalue problems. The main goal of QPE is to estimate the phase (eigenvalue) of an eigenvector of a unitary operator with high precision.

At its core, QPE aims to estimate the phase ϕ in the expression $U|\psi\rangle = e^{2\pi i\phi}|\psi\rangle$, where U is a unitary operator and $|\psi\rangle$ is an eigenvector of U. The phase ϕ lies in the range $[0, 1)$. The algorithm leverages the principles of quantum superposition and interference to extract this phase efficiently.

3.3.1 Steps of the Quantum Phase Estimation Algorithm Steps

The QPE has five main steps on which we describe below.

1. **Initialization**. Prepare two quantum registers. The first register consists of n qubits, all initialized to $|0\rangle$. The second register contains enough qubits to represent the eigenvector $|\psi\rangle$, initialized to the state $|\psi\rangle$. Therefore, the initialization is
$$|\psi\rangle \otimes |0\rangle^{\otimes n} \tag{3.18}$$
2. **Hadamard Transform**. Apply the Hadamard gate H to each of the n qubits in the first register. This creates a superposition of all possible states:
$$\frac{1}{\sqrt{2^n}}\sum_{k=0}^{2^n-1}|k\rangle \tag{3.19}$$
Therefore, the Hadamard Transform is given by
$$\frac{1}{\sqrt{2^n}}\sum_{k=0}^{2^n-1}|k\rangle \otimes |\psi\rangle \tag{3.20}$$
3. **Controlled Unitary Operations**. For each qubit j in the first register, apply a controlled unitary operation U^{2^j} with the j-th qubit as the control qubit and the second register as the target. This operation entangles the phase information

with the state of the first register. The controlled unitary operations are given:

$$\frac{1}{\sqrt{2^n}} \sum_{k=0}^{2^n-1} |k\rangle \otimes U^k |\psi\rangle \tag{3.21}$$

and given $U|\psi\rangle = e^{2\pi i\phi}|\psi\rangle$, this step transforms the state into:

$$\frac{1}{\sqrt{2^n}} \sum_{k=0}^{2^n-1} e^{2\pi ik\phi} |k\rangle \otimes |\psi\rangle \tag{3.22}$$

4. **Inverse Quantum Fourier Transform**. Apply the inverse Quantum Fourier Transform (QFT†) to the first register. The QFT† converts the phase information encoded in the amplitudes of the quantum states into a computational basis state that directly represents the binary fraction of the phase ϕ. The inverse QFT on the first register transforms the phase-encoded state into:

$$|\widetilde{\phi}\rangle \otimes |\psi\rangle \tag{3.23}$$

where $\widetilde{\phi}$ is the binary representation of the estimated phase ϕ.
5. **Measurement**. Measure the first register. The result is a binary string that approximates ϕ to n bits of precision. Upon measuring the first register, the result is an n-bit approximation of ϕ:

$$\phi \approx \frac{m}{2^n} \tag{3.24}$$

where m is the measured integer.

3.3.2 *Implementation*

The implementation for the Quantum Phase Estimation is given in Listing and follows the steps presented in Sect. 3.3.1, while the result is shown in Fig. 3.12.

```
from qiskit_aer import AerSimulator
from qiskit import QuantumCircuit, transpile
from qiskit.circuit.library import QFT
from qiskit.visualization import plot_histogram
import numpy as np

# Define the number of qubits in the counting register (n) and the eigenstate register
n = 3
ancilla_qubits = 1

# Create a Quantum Circuit with n counting qubits and 1 eigenstate qubit
qc = QuantumCircuit(n + ancilla_qubits, n)

# Initialize the eigenstate qubit to |1>
qc.x(n)

# Apply Hadamard gates to the counting qubits
qc.h(range(n))
```

```
    # Apply controlled unitary operations
    # Let's consider U = X (Pauli-X gate) and apply it in controlled fashion
    # The eigenstate of X is |+> and |-> with eigenvalues +1 and -1, respectively.
    # For simplicity, we consider X |1> = |0> with phase 0.5 (i.e., 1/2 in binary)
    for qubit in range(n):
    repetitions = 2 ** qubit
    for _ in range(repetitions):
    qc.cx(qubit, n)

    # Apply the inverse Quantum Fourier Transform
    qc.append(QFT(num_qubits=n, inverse=True, do_swaps=False).decompose(), range(n))

    # Measure the counting qubits
    qc.measure(range(n), range(n))

    # Draw the circuit
    qc.draw(output='mpl')

    # Execute the circuit on the AER simulator
    backend = AerSimulator()
    job = transpile(qc, backend)

    print(job)
```

Listing 3.2 Implementation of the QPE

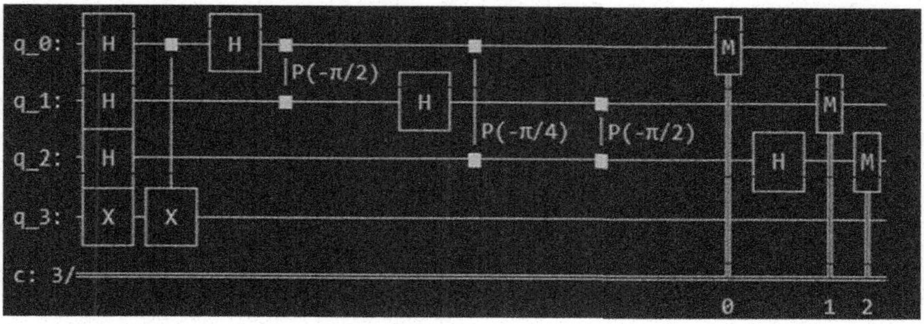

Fig. 3.12 QFT Python output

3.3.3 Performance Analysis

Below we analyze the QPE algorithm considering time complexity, space complexity, and execution time.

Time complexity. The time complexity of the QPE algorithm is determined by the steps of the QPE described previously. Each initialization and Hadamard transform requires $O(n)$ time complexity. In controlled unitary operations, For each qubit j in the counting register, we apply the controlled unitary operation U^{2^j}. The number of gates required for this operation grows exponentially with j, resulting in a complexity of $O(2^n)$ for all controlled operations. Further, Applying the inverse Quantum Fourier Transform to n qubits involves a series of controlled-phase gates and Hadamard gates.

The time complexity of the QFT is $O(n^2)$. Lastly, measuring the qubits takes $O(n)$ time. Combining these steps, the overall time complexity of the QPE algorithm is dominated by the controlled unitary operations, leading to a time complexity of $O(2^n)$.

Execution time of the QPE algorithm can be analyzed by considering several factors: the number of qubits, the complexity of the controlled unitary operations, and the efficiency of the quantum hardware or simulator being used, as follows:

- **Number of qubits**: The execution time increases with the number of qubits n in the counting register. More qubits lead to more complex operations and longer circuit depths.
- **Controlled unitary operations**: The complexity of the controlled unitary operations U^{2^j} exponentially increases with the number of qubits. These operations are the most time-consuming part of the algorithm.
- **Quantum Fourier Transform**: The inverse Quantum Fourier Transform (QFT) has a quadratic complexity in terms of the number of qubits.
- **Quantum Hardware or Simulator**: The execution time depends on the performance of the quantum hardware or the efficiency of the simulator. Quantum hardware may have additional overhead due to gate execution times, decoherence, and noise.

Space complexity. The space complexity of the QPE algorithm is determined by the number of qubits used and the classical bits needed for measurement:

- **Qubits**. The algorithm uses n qubits for the counting register and one additional qubit for the eigenstate register. Thus, the space complexity in terms of qubits is $O(n)$.
- **Classical bits**. To store the measurement results, we need n classical bits. This also contributes $O(n)$ to the space complexity.

3.4 Variational Quantum Eigensolver

The Variational Quantum Eigensolver (VQE) [3] is an algorithm that utilizes the principles of quantum computing to solve for the eigenvalues and eigenstates of a Hamiltonian. It is particularly effective for finding the ground state energy of quantum systems, a problem very important in quantum chemistry and condensed matter physics. The VQE is a hybrid quantum-classical algorithm, combining the strengths of quantum circuits to represent complex wavefunctions and classical optimization techniques to minimize the energy expectation value.

3.4.1 Variational Quantum Eigensolver Steps

The VQE algorithm consists of six main steps, as follows:

1. **Ansatz Selection**: Choose a parameterized quantum circuit (ansatz) to represent the trial wavefunction. The ansatz is a trial wavefunction parameterized by a set of variables. It is represented by a quantum circuit $U(\boldsymbol{\theta})$, where $\boldsymbol{\theta}$ is a vector of parameters. A common choice for the ansatz is the Unitary Coupled Cluster (UCC) ansatz, particularly useful in quantum chemistry.

$$|\psi(\boldsymbol{\theta})\rangle = U(\boldsymbol{\theta})|\psi_0\rangle \tag{3.25}$$

Here, $|\psi_0\rangle$ is an initial reference state, often chosen as a Hartree-Fock state.
2. **Initialization**: Initialize the parameters of the ansatz. Initialize the parameters $\boldsymbol{\theta}$ either randomly or based on heuristic approaches:

$$\boldsymbol{\theta} = \boldsymbol{\theta}_0 \tag{3.26}$$

3. **Quantum State Preparation**: Prepare the quantum state using the parameterized ansatz on a quantum computer:

$$|\psi(\boldsymbol{\theta})\rangle = U(\boldsymbol{\theta})|\psi_0\rangle \tag{3.27}$$

4. **Measurement**: Measure the expectation values of the Hamiltonian components. The Hamiltonian H of the quantum system is expressed as a sum of Pauli operators:

$$H = \sum_i c_i P_i \tag{3.28}$$

where c_i are coefficients and P_i are tensor products of Pauli matrices. The expectation value of the Hamiltonian is:

$$E(\boldsymbol{\theta}) = \langle\psi(\boldsymbol{\theta})|H|\psi(\boldsymbol{\theta})\rangle = \sum_i c_i \langle\psi(\boldsymbol{\theta})|P_i|\psi(\boldsymbol{\theta})\rangle \tag{3.29}$$

To find $E(\boldsymbol{\theta})$, measure the expectation values $\langle\psi(\boldsymbol{\theta})|P_i|\psi(\boldsymbol{\theta})\rangle$ on the quantum computer.
5. **Classical Optimization**: Use a classical optimizer to update the parameters of the ansatz to minimize the energy expectation value. The energy expectation value $E(\boldsymbol{\theta})$ is minimized using a classical optimization algorithm. Popular choices include gradient descent, conjugate gradient, and Nelder-Mead methods. The optimization updates the parameters $\boldsymbol{\theta}$ to reduce the energy:

$$\boldsymbol{\theta} \leftarrow \boldsymbol{\theta} - \eta \nabla_{\boldsymbol{\theta}} E(\boldsymbol{\theta}) \tag{3.30}$$

where η is the learning rate.

```
Minimum eigenvalue: -1.8566052344041526
Optimal parameters:
{ParameterVectorElement(θ[13]): -1.3949896353914588,
 ParameterVectorElement(θ[14]): -4.792443893106996,
 ParameterVectorElement(θ[15]): 1.229551904672718,
 ParameterVectorElement(θ[0]): -2.389374852079809,
 ParameterVectorElement(θ[1]): -0.7135202059065356,
 ParameterVectorElement(θ[2]): 5.966332586415597,
 ParameterVectorElement(θ[3]): 4.579008830872718,
 ParameterVectorElement(θ[4]): 6.170738537212222,
 ParameterVectorElement(θ[5]): -4.442173300338169,
 ParameterVectorElement(θ[6]): 2.053268886468512,
 ParameterVectorElement(θ[7]): 5.556347592738521,
 ParameterVectorElement(θ[8]): -2.7087946273913097,
 ParameterVectorElement(θ[9]): -3.171565612641914,
 ParameterVectorElement(θ[10]): -4.60077622006849,
 ParameterVectorElement(θ[11]): -1.4845809074571306,
 ParameterVectorElement(θ[12]): 3.629511397168642}
```

Fig. 3.13 VQE Python output

6. **Iteration**: Repeat steps 3 to 5 iteratively. The quantum state preparation and measurement are repeated with updated parameters until convergence criteria are met, usually when the change in energy $E(\boldsymbol{\theta})$ between iterations falls below a threshold.

Upon convergence, the parameters $\boldsymbol{\theta}^*$ correspond to the minimum energy found, and the quantum state $|\psi(\boldsymbol{\theta}^*)\rangle$ approximates the ground state of the Hamiltonian.

3.4.2 Implementation

The implementation for the Variational Quantum Solver is given in Listing 3.3 and follows the steps presented in Sect. 3.4.1, while the result is shown in Fig. 3.13.

```
from qiskit_aer import AerSimulator
from qiskit.circuit.library import TwoLocal
from qiskit_algorithms import VQE
from qiskit_algorithms.optimizers import COBYLA
from qiskit.quantum_info import SparsePauliOp
from qiskit.primitives import Estimator
from pprint import pprint

# Define the Hamiltonian for the H2 molecule using SparsePauliOp
pauli_list = [("II", -1.0523732), ("IZ", 0.3979374), ("ZI", -0.3979374),
("ZZ", -0.0112801), ("XX", 0.1809312)]
hamiltonian = SparsePauliOp.from_list(pauli_list)

# Choose an ansatz (variational form)
ansatz = TwoLocal(rotation_blocks=['ry', 'rz'], entanglement_blocks='cz',
```

```
        entanglement='full', reps=3)

    # Choose a classical optimizer
    optimizer = COBYLA(maxiter=100)

    # Get the backend
    backend = AerSimulator()

    # Set up the VQE algorithm using the Estimator primitive
    vqe = VQE(estimator=Estimator(), ansatz=ansatz, optimizer=optimizer)

    # Execute the VQE algorithm to find the ground state energy
    result = vqe.compute_minimum_eigenvalue(operator=hamiltonian)

    # Print the result
    print(f"Minimum eigenvalue: {result.eigenvalue.real}")
    print("Optimal parameters:")
    pprint(result.optimal_parameters)
```

Listing 3.3 Implementation of the Variational Quantum Solver

3.4.3 Performance Analysis

We analyze the performance of the VQE following this metrics: time analysis, space analysis, and execution time.

Time analysis. The time complexity of the VQE algorithm can be broken down into several components:

- **Quantum circuit preparation**. The time required to prepare the quantum state using the ansatz. If the ansatz has the depth d involves n qubits, the time complexity is $O(d \cdot n)$.
- **Measurement**. The number of measurements needed depends on the number of terms in the Hamiltonian, N_H. Each term requires repeated measurements to obtain accurate expectation values. If each term requires M measurements, the time complexity for measurements is $O(N_H \cdot M)$.
- **Classical optimization**. The classical optimizer iteratively updates the parameters to minimize the energy expectation value. Let T be the number of iterations required for convergence, and let k be the number of parameters. The time complexity of each iteration depends on the specific optimization algorithm. For example, the COBYLA optimizer has a time complexity of $O(k)$ per iteration, resulting in a total complexity of $O(T \cdot k)$.

Combining these components, the overall time complexity of the VQE algorithm is approximately $O(T \cdot k \cdot (d \cdot n + N_H \cdot M))$.

The **execution time** of the VQE algorithm depends on the specifics of the quantum hardware and classical optimizer used. Here are some key factors:

- **Quantum circuit execution**: The time to execute the quantum circuit on a quantum computer depends on the gate execution times and the depth d of the circuit. For simulators, this also includes the overhead of simulating the quantum state evolution.

- **Measurement time**: The time required for measurements depends on the number of shots M and the number of Hamiltonian terms N_H. Each measurement involves repeated execution of the quantum circuit, which can be time-consuming on real quantum hardware.
- **Classical optimization time**: The execution time for the classical optimizer depends on the number of iterations T and the complexity of the optimization algorithm. Optimizers like COBYLA are relatively efficient, but the total optimization time can still be significant for a large number of parameters k.

Space complexity. The space complexity of the VQE algorithm can be considered in terms of both quantum and classical resources:

- **Quantum resources**: The primary quantum resource is the number of qubits n. Additionally, quantum circuits require a number of gates proportional to the depth d of the ansatz.
- **Classical resources**: Classical resources include storage for the parameters $\boldsymbol{\theta}$ (with space complexity $O(k)$) and inter mediate results during optimization. The overall space complexity for classical resources is $O(k + N_H)$, where N_H accounts for the storage of Hamiltonian terms.

Therefore, the total space complexity is dominated by the quantum resource requirement $O(n)$ and the classical resource requirement $O(k + N_H)$.

3.5 Shor's Algorithm

Shor's Algorithm [4], developed by Peter Shor in 1994, is a quantum computing algorithm designed for integer factorization. The security of cryptographic systems that depend on the difficulty of integer factorization, such as cryptosystem [5], may be compromised by the significant acceleration of the factorization process by this groundbreaking algorithm.

For large integers, factorization is computationally challenging, fact that makes it ideal for cryptographic systems because this ensures that the difficulty of breaking the encryption is high. Integer factorization is the process through which a composite number is broken down into its prime factors. In order to guarantee security, classical algorithms necessitate superpolynomial time to factor large integers. However, using Shor's Algorithm it is not the case anymore because it provides a polynomial-time solution on a quantum computer.

3.5.1 Shor's Algorithm Steps

Shor's Algorithm has four main steps on which we present below.

Fig. 3.14 Shor's algorithm Python output

```
(2, 11)
```

1. **Quantum Fourier Transform (QFT) Initialization**. In this step the qubits are initialized in a superposition of states using QFT, which is the quantum version of the discrete Fourier transform, and it translates quantum information into a frequency spectrum.
2. **Modular Exponentiation**. The algorithm then uses a function that performs modular exponentiation. Given an integer N (the number to be factored) and a value a with $a < N$ randomly chosen, the exponentiation computation computes $a^x \mod N$, for different values of x. This step is one of the most important because it sets up the conditions for the quantum interference patterns that the QFT will exploit.
3. **Quantum Fourier Transform**. After applying the modular exponentiation, a second QFT is applied. This step transforms the state again and extracts the periodicity information from the quantum state, which is an important component for finding the factors in factorization process.
4. **Measurement and Post-processing**. The output of the QFT is measured. This output is probabilistically related to the period of $a^x \mod N$. Then, using classical computations (instead of quantum computations), the greatest common divisor (GCD) between N and the numbers obtained from period. This leads a factor of N.

An important aspect in Shor's algorithm is finding the order r of $a \mod N$. The order is given by the smallest integer r, such that $a^r \equiv 1(\mod N)$, which is found by applying the quantum Fourier transform. Once r is computed, it is checked whether r is even. If true, then $\gcd(a^{r/2} - 1, N)$ and $\gcd(a^{r/2} + 1, N)$ are likely non-trivial factors of N.

3.5.2 Implementation

Implementation of Shor's Algorithm involves following the steps from Sect. 3.5.1, including setting up the quantum circuit, applying quantum Fourier transform, and performing classical post-processing. For this implementation we used Cirq library from Python. The code is presented in Listing 3.4 and, while the result is shown in Fig. 3.14.

```
import cirq
import numpy as np
from fractions import Fraction
import math
from typing import List, Tuple, Optional

def qft(qubits):
circuit = cirq.Circuit()
n = len(qubits)
for i in range(n):
```

```
circuit.append(cirq.H(qubits[i]))
for j in range(i + 1, n):
circuit.append(cirq.CZ(qubits[j], qubits[i]) ** (1 / 2**(j - i)))
return circuit
def inverse_qft(qubits):
circuit = cirq.Circuit()
n = len(qubits)
for i in range(n):
for j in range(i):
circuit.append(cirq.CZ(qubits[i], qubits[j]) ** (-1 / 2**(i - j)))
circuit.append(cirq.H(qubits[i]))
return circuit
def continued_fraction_expansion(x: float) -> List[int]:
expansion = []
while x:
a = int(x)
expansion.append(a)
x -= a
if x == 0:
break
x = 1 / x
return expansion
def find_period(measurement: int, num_qubits: int, x: int, N: int) -> Optional[int]:
frac = measurement / (2 ** num_qubits)
expansion = continued_fraction_expansion(frac)
for i in range(1, len(expansion)):
denom = Fraction(0, 1)
for term in reversed(expansion[:i]):
denom = 1 / (term + denom)
r = denom.denominator
if r % 2 == 0 and pow(x, r // 2, N) != N - 1:
return r
return None
def shor_circuit(x: int, N: int) -> Tuple[cirq.Circuit, List[cirq.LineQubit]]:
num_qubits = math.ceil(np.log2(N)) * 2
qubits = cirq.LineQubit.range(num_qubits)
circuit = cirq.Circuit()
# Prepare superposition
circuit.append(cirq.H.on_each(*qubits[:num_qubits // 2]))
# Modular exponentiation
for i in range(num_qubits // 2):
exp = 2 ** i
for _ in range(exp):
for j in range(num_qubits // 2, num_qubits):
circuit.append(cirq.CZ(qubits[i], qubits[j]))
# Apply inverse QFT
circuit += inverse_qft(qubits[:num_qubits // 2])
# Measurement
circuit.append(cirq.measure(*qubits[:num_qubits // 2], key='result'))
return circuit, qubits

def shor_algorithm(N: int) -> Optional[Tuple[int, int]]:
if N % 2 == 0:
return 2, N // 2
for attempt in range(5):
x = np.random.randint(2, N)
if math.gcd(x, N) != 1:
return math.gcd(x, N), N // math.gcd(x, N)

circuit, qubits = shor_circuit(x, N)
simulator = cirq.Simulator()
result = simulator.run(circuit)
measurements = result.measurements['result'][0]
measurement = 0
for bit in measurements[::-1]:
measurement = (measurement << 1) | bit

num_qubits = len(qubits) // 2
r = find_period(measurement, num_qubits, x, N)
if r is None:
continue

if r % 2 != 0:
continue

factor1 = math.gcd(pow(x, r // 2) - 1, N)
```

```
factor2 = math.gcd(pow(x, r // 2) + 1, N)

if factor1 in [1, N] or factor2 in [1, N]:
continue

return factor1, factor2

return None

N = 22  # Number to factorize
factors = shor_algorithm(N)
print(factors)
```

Listing 3.4 Implementation of Shor's algorithm

3.5.3 Performance Analysis

Shor's Algorithm, significant for its use in cryptography, provides a quantum approach to factorizing large integers efficiently. Analyzing its performance involves considering theoretical aspects such as time complexity and space complexity, along with practical metrics such as execution time on actual or simulated quantum hardware.

Time Complexity. Shor's Algorithm brings a high degree of time performance for integer factorization compared to classical approach, where the best known algorithm is () [6] whose complexity is $O(\exp((\log N)^{\frac{1}{3}}(\log\log N)^{\frac{2}{3}}))$, where N is the integer to be factorized. Comparatively, Shor's Algorithm has a polynomial-time solution:

- **Quantum part**: for finding the period r, therefore the quantum computations, has a complexity of $O((\log N)^2(\log\log N)(\log\log\log N))$ due to the quantum Fourier transform and quantum arithmetic operations.
- **Overall time complexity**: When including both quantum and classical parts, the overall time complexity remains polynomial $O(\log N)$. Thus, it is exponentially faster than any classical algorithm for a large integer N.

However, the execution time on quantum hardware or simulators can drastically vary due to several factors:

- **Hardware limitations**: current quantum devices, known as *Noisy Intermediate-Scale Quantum* (NISQ) devices, have limitations such as gate fidelity, qubit coherence times, and error rates that can affect performance.
- **Simulator performance**: on classical simulators, the execution time is influenced by the classical computational resources available (such as CPU speed, memory), as simulators need to manage the exponential growth in state complexity as more qubits are simulated.
- **Example on Qiskit simulator**: running Shor's Algorithm to factorize small numbers might take only a few seconds on a simulator, but these values for time complexity are not representative of how the algorithm would perform on actual

quantum hardware with larger numbers. Factors like initialization, gate application, and measurement contribute to these times but do not scale linearly with increased problem size.

Space Complexity. The space complexity of Shor's Algorithm depends on the number of qubits required to represent the numbers involved and to perform the required quantum operations:

- **Qubit Requirements**: usually, the algorithm needs $2 \cdot \log N + O(1)$ qubits, as follows: one register of $\log N$ qubits are used to hold the large integer N and another $\log N$ quibits are used to compute modular exponentiation. Another several qubits are used to carry out the quantum Fourier transform and other required operations.
- **Overall Space Complexity**: this leads to a linear space complexity in terms of the number of qubits as a function of the input size's logarithm. This is highly space-efficient compared to many classical algorithms that may require more extensive memory allocation for handling large integers and their factorizations.

3.6 Grover's Algorithm

Grover's algorithm [7], proposed by Lov Grover in 1996, is another important contribution in the field of quantum computing, demonstrating how quantum mechanics has a better performance compared to classical methods in certain computational tasks. This algorithm provides a quadratic speedup for unstructured search problems, where a specific item within an unsorted database should be found. While classical computations require $O(N)$ operations, Grover's algorithm requires $O(\sqrt{N})$ operations, proving the significance of quantum parallelism.

3.6.1 Grover's Algorithm Steps

Shor's Algorithm has five main steps on which we present below.

1. **Initialization**. The algorithm starts by initializing an n-qubit system to the uniform superposition of all possible states. Initially, the system is in the state $|0\rangle^{\otimes n}$. Applying the Hadamard gate to each qubit transforms this into a uniform superposition:

$$H^{\otimes n}|0\rangle^{\otimes n} = \frac{1}{\sqrt{N}} \sum_{x=0}^{N-1} |x\rangle, \tag{3.31}$$

   where $N = 2^n$ is the total number of states.
2. **Oracle Query**. An oracle function O marks the correct state $|w\rangle$ by inverting its amplitude. It is defined as follows:

$$O|x\rangle = \begin{cases} -|x\rangle, \text{ if } x = w \\ |x\rangle, \text{ if } x \neq w \end{cases} \tag{3.32}$$

This operation inverts the amplitude of the target state while leaving other states unchanged.

3. **Amplitude Amplification**. This involves to apply the Grover diffusion operator G, representing the core of this algorithm, which amplifies the probability amplitude of the target state. It is computed as follows:

$$G = H^{\otimes n}(2|0\rangle\langle 0| - I)H^{\otimes n}, \tag{3.33}$$

where I is the identity matrix. The combined effect of the oracle and the diffusion operator after each iteration is a rotation in the two-dimensional space spanned by $|w\rangle$ and $\frac{1}{N-1}\sum_{x\neq w}|x\rangle$.

4. **Iteration**. The oracle query and amplitude amplification, i.e. Grover's diffusion operator, are repeated approximately by $\sqrt{N}$ times. Each iteration increases the probability amplitude of the target state. The exact number of iterations, r, is given by

$$r \approx \frac{\pi}{4}\sqrt{N} \tag{3.34}$$

This ensures that the amplitude of the target state is maximized, making it highly likely to be measured.

5. **Measurement**. After the iterations, the quantum state is measured. The measurement collapses the superposition to one of the basis states, with the target state $|w\rangle$ having a high probability.

3.6.2 Implementation

The implementation for the Grover's algorithm is given in Listing 3.5 and follows the steps presented in Sect. 3.6.1, while the result is shown in Fig. 3.15.

```
import numpy as np
from qiskit_aer import AerSimulator
from qiskit import QuantumCircuit, transpile, assemble
from qiskit.visualization import plot_histogram

# Oracle for marking the state |11>
def grover_oracle(nqubits):
oracle = QuantumCircuit(nqubits)
oracle.cz(nqubits-2, nqubits-1)
return oracle.to_gate(label="Oracle")

# Grover's diffusion operator
def diffusion_operator(nqubits):
diffuser = QuantumCircuit(nqubits)
diffuser.h(range(nqubits))
diffuser.x(range(nqubits))
diffuser.h(nqubits-1)
diffuser.mcx(list(range(nqubits-1)), nqubits-1)
diffuser.h(nqubits-1)
diffuser.x(range(nqubits))
```

```
diffuser.h(range(nqubits))
return diffuser.to_gate(label="Diffusion")

# Function to create Grover's algorithm circuit
def grover_circuit(nqubits):
qc = QuantumCircuit(nqubits, nqubits)

# Step 1: Apply Hadamard gates to all qubits
qc.h(range(nqubits))

# Step 2: Apply the oracle and diffusion operators
iterations = int(np.floor(np.pi / 4 * np.sqrt(2**nqubits)))  # Number of iterations
for _ in range(iterations):
qc.append(grover_oracle(nqubits), range(nqubits))
qc.append(diffusion_operator(nqubits), range(nqubits))

# Step 3: Measure all qubits
qc.measure(range(nqubits), range(nqubits))
return qc

# Define the number of qubits
nqubits = 2

# Create and execute the Grover's algorithm circuit
qc = grover_circuit(nqubits)
simulator = AerSimulator()
compiled_circuit = transpile(qc, simulator)
qobj = simulator.run(compiled_circuit, shots=1)
result = qobj.result()
counts = result.get_counts()

# Plot the results
print("Resulting counts:", counts)
plot_histogram(counts)
```

Listing 3.5 Implementation of the Grover's algorithm

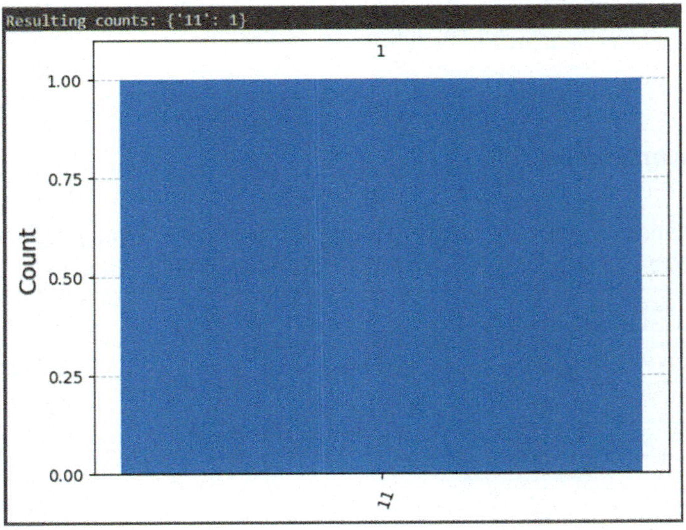

Fig. 3.15 Grover's algorithm Python output

3.6.3 Performance Analysis

Grover's Algorithm provides a quadratic speedup for unstructured search problems, making it significantly faster than classical search algorithms.

Time complexity. The quantum time complexity is characterized by the number of iterations and the operations within each iteration, leading to a total complexity of $O(n \log n)$, as follows:

- **Number of iterations**: The core of Grover's Algorithm lies in the number of iterations required to amplify the amplitude of the desired state. The optimal number of iterations is approximately $\frac{\pi}{4}\sqrt{N}$, where N is the number of elements in the search space.
- **Each iteration**: Each iteration consists of applying the oracle and the diffusion operator, both of which operate on all qubits and require $O(n)$ operations for n qubits.

Compared to a classical linear search with time complexity $O(N)$, Grover's Algorithm provides a quadratic speedup.
Execution time is an empirical measure that depends on the specifics of the quantum hardware and the efficiency of the implementation. Several factors affect the execution time:

- **Gate times**: the time to perform individual quantum gates, such as Hadamard gates, X gates, and controlled-Z gates. These times vary but usually range from nanoseconds to microseconds per gate.
- **Number of gates**: Each iteration involves a fixed sequence of gates. Since we need $O(\sqrt{N})$ iterations and each involves $O(n)$ gates, the total number of gate operations scales as $O(n\sqrt{N})$.
- **Quantum hardware limitations**: Current quantum hardware has limitations in terms of coherence time, gate fidelity, and error rates. These factors need error correction or multiple trials, impacting the practical execution time.
- **Simulator vs. real hardware**: On quantum simulators, execution time is limited by classical computational power. On real quantum hardware, it is affected by initialization, gate execution, and readout times.

Space complexity. The space complexity of Grover's Algorithm is primarily determined by the number of qubits and classical bits used in the implementation:

- **Quantum space**: to represent a database with N elements, n qubits are needed, where $N = 2^n$. Therefore, the quantum space is $O(n)$ qubits.
- **Classical space**: Also, n classical bits are needed to store the measurement results from the quantum register.

Therefore, the overall space complexity is $O(n)$ qubits and $O(n)$ classical bits.

3.7 Deutsch–Jozsa Algorithm

The Deutsch–Jozsa algorithm [8] is a pioneering quantum algorithm that exemplifies the power of quantum computation over classical methods. Proposed by David Deutsch and Richard Jozsa in 1992, it determines if a black-box function f is constant or balanced exponentially faster than any deterministic classical algorithm.

The problem addressed by the Deutsch–Jozsa algorithm involves a black-box function $f : \{0, 1\}^n \rightarrow \{0, 1\}$. The function is either constant (outputting the same value for all inputs) or balanced (outputting 0 for exactly half of the inputs and 1 for the other half). The task is to determine whether f is constant or balanced.

In classical computation, determining whether f is constant or balanced would require evaluating f at least $2^{n-1} + 1$ times in the worst case. However, the Deutsch–Jozsa algorithm accomplishes this with a single evaluation of f using a quantum computer.

3.7.1 *Deutsch–Jozsa Algorithm Steps*

The Deutsch–Jozsa algorithm involves the following steps:

1. **Initialization**. Prepare $n + 1$ qubits in the state $|0\rangle^{\otimes n}|1\rangle$. This means n qubits are in the state $|0\rangle$ and one qubit is in the state $|1\rangle$.
2. **Apply Hadamard Gates to each qubit**. The state transforms to:

$$\frac{1}{\sqrt{2^n}} \sum_{x=0}^{2^n-1} |x\rangle \left(\frac{|0\rangle - |1\rangle}{\sqrt{2}} \right) \tag{3.35}$$

   Simplifying, this becomes:

$$\frac{1}{\sqrt{2^{n+1}}} \sum_{x=0}^{2^n-1} |x\rangle \big(|0\rangle - |1\rangle \big) \tag{3.36}$$

3. **Oracle application**. Apply the oracle U_f. The state changes to

$$\frac{1}{\sqrt{2^{n+1}}} \sum_{x=0}^{2^n-1} (-1)^{f(x)} |x\rangle \big(|0\rangle - |1\rangle \big) \tag{3.37}$$

4. **Apply the Hadamard gate** to the first n qubits again. This operation can be expressed as:

$$H^{\otimes n} |x\rangle = \frac{1}{\sqrt{2^n}} \sum_{z=0}^{2^n-1} (-1)^{x \cdot z} |z\rangle \tag{3.38}$$

Therefore, the state becomes:

$$\frac{1}{2^{n+1}}\sum_{x=0}^{2^n-1}\sum_{z=0}^{2^n-1}(-1)^{f(x)}(-1)^{x\cdot z}|z\rangle(|0\rangle-|1\rangle) \tag{3.39}$$

5. **Measurement**. Measure the first n qubits. If the function f is constant, the measurement will always yield to $|0\rangle^{\otimes n}$. If f is balanced, the probability of measuring $|0\rangle^{\otimes n}$ is zero, showing that f is balanced.

To understand why the algorithm works, consider the final state before measurement:

$$\frac{1}{2^{n+1}}\sum_{z=0}^{2^n-1}\left[\sum_{x=0}^{2^n-1}(-1)^{f(x)\oplus(x\cdot z)}\right]|z\rangle(|0\rangle-|1\rangle) \tag{3.40}$$

- Constant f: if f is constant, $(-1)^{f(x)}$ is the same for all x, and the inner sum simplifies to 2^n for $z=0$ and 0 for all $z\neq 0$. Thus, measuring the first n qubits yields $|0\rangle^{\otimes n}$.
- Balanced f: if f is balanced, the sum $(-1)^{f(x)\oplus(x\cdot z)}$ over all x is zero for $z=0$ and non-zero for other values of z. Therefore, the probability of measuring $|0\rangle^{\otimes n}$ is zero, showing that f is balanced.

3.7.2 Implementation

The implementation for the Deutsch-Jozsa algorithm is given in Listing 3.6 and follows the steps presented in Sect. 3.7.1, while the result is shown in Fig. 3.16.

```
from qiskit_aer import AerSimulator
from qiskit import QuantumCircuit, transpile, assemble
from qiskit.visualization import plot_histogram

def deutsch_jozsa_algorithm(n, oracle):

# Create a quantum circuit with n+1 qubits and n classical bits
qc = QuantumCircuit(n + 1, n)
# Apply Hadamard gates to the first n qubits
qc.h(range(n))
# Prepare the ancillary qubit in the |1> state and apply Hadamard gate
qc.x(n)
qc.h(n)

# Append the oracle
qc.append(oracle, range(n + 1))

# Apply Hadamard gates to the first n qubits again
qc.h(range(n))

# Measure the first n qubits
qc.measure(range(n), range(n))

# Execute the circuit on a simulator
simulator = AerSimulator()
compiled_circuit = transpile(qc, simulator)
```

```
qobj = simulator.run(compiled_circuit, shots=1)
result = qobj.result()

# Get the counts (measurement results)
counts = result.get_counts()
return counts

# Example: Creating an oracle for a balanced function f(x) = x0 XOR x1
def create_balanced_oracle():
oracle = QuantumCircuit(3)  # 2 input qubits + 1 ancillary qubit

# Apply CNOT gates to flip the ancillary qubit based on the input qubits
oracle.cx(0, 2)
oracle.cx(1, 2)

return oracle

# Example: Creating an oracle for a constant function f(x) = 0
def create_constant_oracle():
oracle = QuantumCircuit(3)  # 2 input qubits + 1 ancillary qubit
# No need to change the ancillary qubit for a constant function
return oracle
# Example usage
n = 2  # Number of qubits (excluding ancillary qubit)
oracle1 = create_balanced_oracle()  # Use a balanced oracle
oracle2 = create_constant_oracle()  # Uncomment to use a constant oracle

result1 = deutsch_jozsa_algorithm(n, oracle1)
result2 = deutsch_jozsa_algorithm(n, oracle2)
print(result1)
print(result1)
```

Listing 3.6 Implementation of the Deutsch-Jozsa algorithm

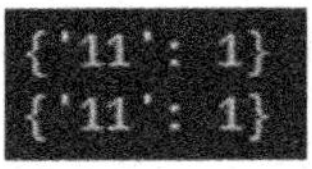

Fig. 3.16 Deutsch-Jozsa algorithm Python output

3.7.3 Performance Analysis

This section presents the time and space complexity of the Deutsch–Jozsa algorithm.

Time complexity. The time complexity of the Deutsch–Jozsa algorithm can be analyzed by considering the number of operations required to determine if a function f is constant or balanced. The Deutsch–Jozsa algorithm requires only a single evaluation of the oracle function, regardless of the number of qubits n, therefore, the time complexity is $O(1)$. For a classical algorithms, the worst case for evaluation of f is given by a complexity of $O(2^n)$.

Space complexity. Space complexity refers to the amount of memory used by the algorithm. For both the classical and quantum versions, we need to consider the number of bits or qubits used:

- **Classical algorithm**: requires space to store the input bits, the function f, and any additional variables used for computation. The space complexity is $O(n)$.

- **Quantum algorithm**: requires $n + 1$ qubits, where n is the number of input qubits and 1 ancillary qubit. Therefore, the space complexity is $O(n)$. Additionally, quantum circuits might need ancillary qubits for intermediate operations, but these do not change the asymptotic space complexity.

3.8 Bernstein-Vazirani Algorithm

The Bernstein-Vazirani algorithm [9] is a quantum computing algorithm that shows the power of quantum computation over classical methods. It was introduced by Ethan Bernstein and Umesh Vazirani in 1992 as part of their work on quantum complexity theory. This algorithm demonstrates how quantum computers can solve certain problems more efficiently than classical computers. The algorithm's primary purpose is to determine a hidden binary string with fewer queries than classical methods.

3.8.1 Bernstein-Vazirani Algorithm Steps

The Bernstein-Vazirani algorithm aims to determine an unknown n-bit binary string $s = s_0 s_1 \ldots s_{n-1}$. Given a function $f_s(x) = s \cdot x$ (where $\cdot$ represents the bitwise dot product modulo 2), the task is to identify the string s. n classical computing, one would need to query the function f_s, multiple times, at least n times, to reconstruct s. The Bernstein-Vazirani algorithm can accomplish this in a single query using quantum principles. The algorithm has five main steps, as follows:

1. **Initialization**: Begin with $n + 1$ qubits in the state $|0\rangle$ and an auxiliary qubit in the state $|0\rangle$:
$$|0\rangle^{\otimes n}|1\rangle \tag{3.41}$$

2. **Hadamard Transformation**: Apply the Hadamard gate (H) to each of the $n + 1$ The Hadamard gate transforms a qubit as follows:
$$H|0\rangle = \frac{1}{\sqrt{2}}(|0\rangle + |1\rangle), H|1\rangle = \frac{1}{\sqrt{2}}(|0\rangle - |1\rangle) \tag{3.42}$$
After applying the Hadamard gate to all qubits, the state of the system becomes:
$$\frac{1}{\sqrt{2^n}} \sum_{x=0}^{2^n-1} |x\rangle \otimes \frac{1}{\sqrt{2}}(|0\rangle - |1\rangle) \tag{3.43}$$

3. **Oracle query**: The function f_s is implemented as an oracle U_f that maps $|x\rangle|y\rangle$ to $|x\rangle|y \oplus f_s(x)\rangle$, where $\oplus$ is the XOR operation. Applying this oracle to the current state:
$$U_f\Big(\frac{1}{\sqrt{2^n}}\sum_{x=0}^{2^n-1}|x\rangle \otimes \frac{1}{\sqrt{2}}(|0\rangle - |1\rangle)\Big) \tag{3.44}$$
Given $f_s(x) = s \cdot x$, the state becomes:
$$\frac{1}{\sqrt{2^n}}\sum_{x=0}^{2^n-1}(-1)^{s \cdot x}|x\rangle \otimes \frac{1}{\sqrt{2}}(|0\rangle - |1\rangle) \tag{3.45}$$
4. **Inverse Hadamard transformation**: Apply the Hadamard transformation again to the first n qubits:
$$H^{\otimes n}\Big(\frac{1}{\sqrt{2^n}}\sum_{x=0}^{2^n-1}(-1)^{s \cdot x}|x\rangle\Big) \tag{3.46}$$
The, using the property of Hadamard transformations and the fact that $H^{\otimes n}|x\rangle = \frac{1}{\sqrt{2^n}}\sum_{z=0}^{2^n-1}(-1)^{x \cdot z}|z\rangle$, the state simplifies to $|s\rangle$.
5. **Measure the first n qubits**. The measurement collapses the state $|s\rangle$ to the binary string s, revealing the hidden string.

The efficiency of the Bernstein-Vazirani algorithm comes from the quantum parallelism and the specific properties of the Hadamard gate. The initial Hadamard transformation creates a superposition of all possible input states. The oracle encodes the hidden string s into the phase of the superposition, while the final Hadamard transformation effectively decodes this phase information into the basis states. Therefore, the hidden string is extracted directly through a single measurement. In mathematical terms, the process is based on the properties of the Hadamard matrix H and the interference patterns resulting from the phase encoding by the oracle. The key insight is that the Hadamard transform of the dot product $s \cdot x$ produces a delta function peaked at the binary string s, allowing the algorithm to isolate s in a single shot.

3.8.2 Implementation

The implementation for the Bernstein-Vazirani algorithm is given in Listing 3.7 and follows the steps presented in Sect. 3.8.1, while the result is shown in Fig. 3.17.

```
from qiskit_aer import AerSimulator
from qiskit import QuantumCircuit, transpile
from qiskit.visualization import plot_histogram
def bernstein_vazirani(s):
n = len(s)  # Number of qubits
qc = QuantumCircuit(n + 1, n)  # n data qubits and 1 ancilla qubit
# Initialize the ancilla qubit in state |1>
qc.x(n)
qc.h(n)
```

```
# Apply Hadamard gates to all data qubits
qc.h(range(n))
# Implement the oracle function
for i in range(n):
if s[i] == '1':
qc.cx(i, n)
# Apply Hadamard gates to all data qubits again
qc.h(range(n))
# Measurement
qc.measure(range(n), range(n))
# Simulate the circuit
simulator = AerSimulator()
compiled_circuit = transpile(qc, simulator)
qobj = simulator.run(compiled_circuit, shots=1)
result = qobj.result()
counts = result.get_counts(qc)
return counts

# Define the hidden string s
s = "1011"  # Example hidden string
# Run the algorithm
counts = bernstein_vazirani(s)
# Print the result
print("Measurement results:", counts)
# Plot the result
plot_histogram(counts)
```

Listing 3.7 Implementation of the Bernstein-Vazirani algorithm

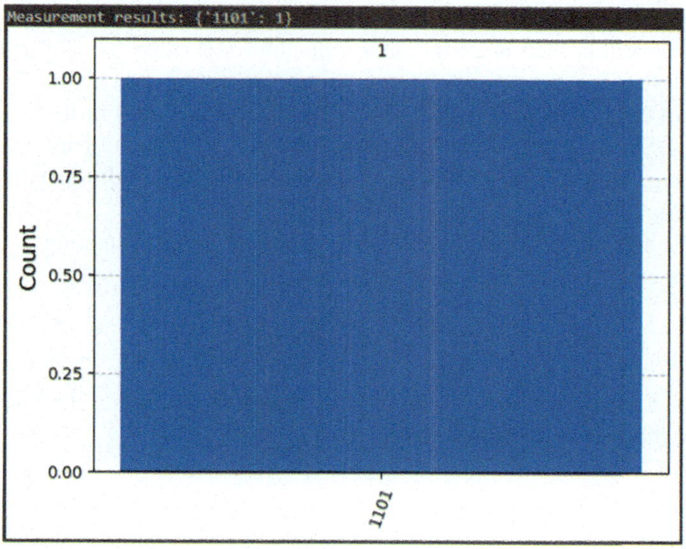

Fig. 3.17 Bernstein-Vazirani algorithm Python output

3.8.3 *Performance Analysis*

To analyze the performance of the Bernstein-Vazirani algorithm, we consider three main metrics: time complexity, space complexity, and execution time.

Time complexity. The time complexity is given by steps presented previously. To initialize $n + 1$ qubits the time is takes is $O(n)$, as well as applying Hadamard gates to all $n + 1$ qubits. The oracle U_f involves n CNOT operations, therefore, the required time is also $O(n)$. The second application of Hadamard gates also takes $O(n)$. Lastly, measuring n qubits takes $O(n)$. Therefore, the overall time complexity is $O(n)$, which is a better performance than the classical version of finding the n-bit hidden string with the overall performance of $O(n^2)$.

The **execution time** for quantum algorithms can be affected by factors such as quantum gate operations, decoherence times, and overhead from error correction if applicable. However, in the context of an ideal quantum simulator or a noise-free quantum computer, the quantum algorithm can be executed much faster than its classical version.

Space complexity. The quantum circuit needs $n + 1$ qubits, and for the result of measurement is needed a classical register of n bits. Therefore, the required quantum space is $O(n)$.

3.9 Simon's Algorithm

Simon's Algorithm [10] is an important quantum algorithm developed by Daniel Simon in 1994. It was one of the first quantum algorithms that shows the potential of quantum computers to solve certain problems exponentially faster than classical computers. Simon's Algorithm addresses the specific problem of finding a hidden binary string, known as the *period*, that satisfies a particular condition within a black-box function.

Simon's algorithm solves the following problem: given a function $f : \{0, 1\} \rightarrow \{0, 1\}^n$ that satisfies the following implication

$$f(x) = f(y) \Leftrightarrow x \oplus y = s \tag{3.47}$$

for a hidden string $s \in \{0, 1\}^n$ and $x \neq y$, where $\oplus$ represents the bitwise XOR operation, the goal is to determine the string s.

3.9.1 Simon's Algorithm Steps

Simon's algorithm has seven main steps on which we present below.

1. **Initialization**. Prepare two quantum registers. The first register is an n-qubit register initialized to the uniform superposition state, and the second register is also an n-qubit register initialized to the $|0\rangle$ state.
2. **Superposition**. Apply Hadamard gates to each qubit in the first register to create an equal superposition of all possible input states:

$$\frac{1}{\sqrt{2^n}} \sum_{x \in \{0,1\}} |x\rangle|0\rangle. \tag{3.48}$$

3. **Oracle query**. Apply the oracle U_f, which maps $|x\rangle|0\rangle$ to $|x\rangle|f(x)\rangle$. This transforms the state to:

$$\frac{1}{\sqrt{2^n}} \sum_{x \in \{0,1\}^n} |x\rangle|f(x)\rangle. \tag{3.49}$$

4. **Measurement of the second register**. Measure the second register. The measurement collapses the state into one of the possible values of $f(x)$, say $f(x_0)$, and the first register into an equal superposition of all x such that $f(x) = f(x_0)$. Due to the promise of the function, this superposition is over x_0 and $x_0 \oplus s$:

$$\frac{1}{\sqrt{2}}\big(|x_0\rangle + |x_0 \oplus s\rangle\big). \tag{3.50}$$

5. **Hadamard Transformation**. For a state $|x\rangle$, the Hadamard transform is applied to each bit of $|x\rangle$:

$$H^{\otimes n}|x\rangle = \frac{1}{\sqrt{2^n}} \sum_{x \in \{0,1\}^n} (-1)^{x \cdot z}|z\rangle, \tag{3.51}$$

where $x \cdot z$ is the dot product between x and z.
6. **Measurement of first register**. The resulting state after applying the Hadamard transform to $\frac{1}{\sqrt{2}}(|x_0\rangle + |x_0 \oplus s\rangle)$ is:

$$\frac{1}{\sqrt{2^{n+1}}} \sum_{z \in \{0,1\}^n} \left((-1)^{x_0 \cdot z} + (-1)^{(x_0 \oplus s)\cdot z}\right)|z\rangle. \tag{3.52}$$

Simplifying the exponent, the state becomes:

$$\frac{1}{\sqrt{2^{n+1}}} \sum_{z \in \{0,1\}^n} (1 + (-1)^{s \cdot z})(-1)^{x_0 \cdot z}|z\rangle. \tag{3.53}$$

This implies that for the terms to be non-zero, $s \cdot z$ must be 0, meaning z must be orthogonal to s.
7. **Collect** $n - 1$ linearly independent measurements $z_1, \ldots, z_{n-1}$ such that $s \cdot z_i = 0$. These measurements form a system of linear equations in modulo 2 arithmetic. Solving this system yields the hidden string s.

Simon's Algorithm is based on the quantum parallelism and interference inherent from quantum mechanics. The most important step is the transformation of the problem into a linear algebraic form that classical algorithms would require exponentially more time to solve. The Hadamard transform creates an interference pattern that encodes information about the hidden string s into measurable quantum states.

This process is exponentially faster than any known classical algorithm for the same problem, proving the potential of quantum computing.

3.9.2 Implementation

The implementation for the Simon's algorithm is given in Listing 3.8 and follows the steps presented in Sect. 3.9.1, while the result is shown in Fig. 3.18.

```
import numpy as np
from qiskit_aer import AerSimulator
from qiskit import QuantumCircuit, transpile, assemble
from qiskit.visualization import plot_histogram
import matplotlib.pyplot as plt
# Define the oracle for Simon's problem
def simon_oracle(s):
n = len(s)
oracle = QuantumCircuit(n * 2)
for qubit in range(n):
if s[qubit] == '1':
oracle.cx(qubit, qubit + n)
return oracle.to_gate(label="Oracle")
# Define the secret string s
s = '101'
# Create the quantum circuit for Simon's algorithm
n = len(s)
circ = QuantumCircuit(n * 2, n)
# Apply Hadamard gates to the first n qubits
circ.h(range(n))
# Append the oracle for the specific problem instance
circ.append(simon_oracle(s), range(n * 2))
# Apply Hadamard gates to the first n qubits again
circ.h(range(n))
# Measure the first n qubits
circ.measure(range(n), range(n))
# Simulate the circuit
simulator = AerSimulator()
transpiled_circ = transpile(circ, simulator)
qobj = simulator.run(transpiled_circ, shots=1)
result = qobj.result()
print(transpiled_circ)
# Get the measurement results
counts = result.get_counts(circ)
print("Measurement_Results:", counts)
# Plot the results
plot_histogram(counts)
plt.show()
```

Listing 3.8 Implementation of the Simon's algorithm

3.9.3 Performance Analysis

Analyzing the performance of Simon's algorithm involves understanding its time complexity and space complexity.

Time complexity. Simon's algorithm demonstrates a significant improvement over classical algorithms for the same problem. The time complexity is divided in:

- **Quantum Query Complexity**: The number of oracle queries made by Simon's Algorithm is $O(n)$, where n is the number of bits in the input. This is because the

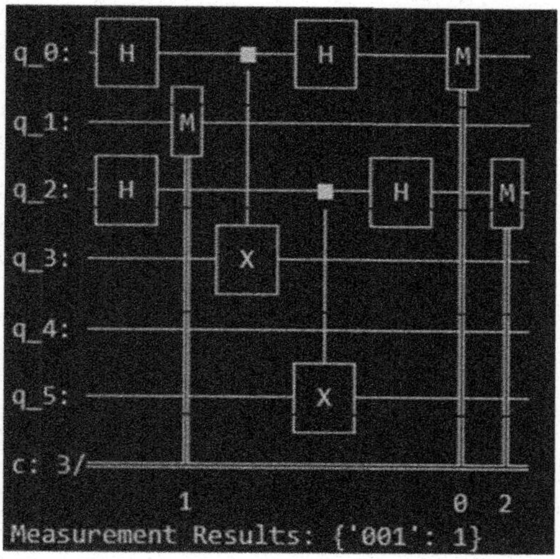

Fig. 3.18 Simon's algorithm Python output

algorithm requires $O(n)$ queries to obtain enough linearly independent equations to solve for the hidden string s.

- **Classical Query Complexity**: A classical algorithm would need $O(2^{n/2})$ queries to solve the same problem. This exponential improvement is a hallmark of the efficiency of Simon's Algorithm.

Space complexity. It is divided in two components, as follows:

- **Quantum space complexity**: Simon's Algorithm uses $2n$ qubits for the computation. This includes n qubits for the computation. This includes n qubits for the output register. Thus, the space complexity is $O(n)$.
- **Classical space complexity**: A classical algorithm solving this problem would require $O(n \cdot 2^n)$ space to store the exponential number of function evaluations and possible input-output pairs.

References

1. Coppersmith, D.: An approximate Fourier transform useful in quantum factoring. arXiv, Jan. 16, 2002. [Online]. http://arxiv.org/abs/quant-ph/0201067
2. Kitaev, A.Y.: Quantum measurements and the Abelian Stabilizer Problem. arXiv, Nov. 20, 1995. [Online]. http://arxiv.org/abs/quant-ph/9511026
3. Peruzzo, A., et al.: A variational eigenvalue solver on a photonic quantum processor. Nat. Commun. **5**(1), 4213 (2014)
4. Shor, P.W.: Algorithms for quantum computation: discrete logarithms and factoring. In: Proceedings 35th Annual Symposium on Foundations of Computer Science, IEEE (1994), pp. 124–134. [Online]. https://ieeexplore.ieee.org/abstract/document/365700/

5. Rivest, R.L., Shamir, A., Adleman, L.: A method for obtaining digital signatures and public-key cryptosystems. Commun. ACM **21**(2), 120–126 (1978). https://doi.org/10.1145/359340.359342
6. Qualls, W.R., Watkins, J., Radtke, D.: A tale of two sieves. In: Proceedings of the International Conference GASTECH (2011). [Online]. http://static.conocophillips.com/files/resources/a-tale-of-two-sieves.pdf
7. Grover, L.K.: A fast quantum mechanical algorithm for database search. In: Proceedings of the Twenty-Eighth Annual ACM Symposium on Theory of Computing - STOC '96, Philadelphia, Pennsylvania, United States. ACM Press (1996), pp. 212–219. https://doi.org/10.1145/237814.237866
8. Deutsch, D., Jozsa, R.: Rapid solution of problems by quantum computation. Proc. R. Soc. Lond. A **439**(1907), 553–558 (1992). https://doi.org/10.1098/rspa.1992.0167
9. Bernstein, E., Vazirani, U.: Quantum complexity theory. In: Proceedings of the Twenty-Fifth Annual ACM Symposium on Theory of Computing - STOC '93, San Diego, California, United States. ACM Press (1993), pp. 11–20. https://doi.org/10.1145/167088.167097
10. Simon, D.R.: On the power of quantum computation. SIAM J. Comput. **26**(5), 1474–1483 (1997). https://doi.org/10.1137/S0097539796298637

Chapter 4
Quantum Cryptography

Quantum cryptography is an approach to securing communication through the principles of quantum mechanics, fundamentally different from traditional cryptographic methods which are based on mathematical complexity. Here's a beginner-friendly introduction to this fascinating field.

At its core, quantum cryptography uses properties of quantum physics to perform cryptographic tasks, one of the most prominent being the secure distribution of encryption keys. Unlike classical cryptography, where the security is typically based on the assumed difficulty of mathematical problems, quantum cryptography's security is based on the laws of physics, making it potentially unbreakable.

In this scenario, the Hadamard transformation (H) is employed to place a qubit ($\begin{bmatrix} 1 \\ 0 \end{bmatrix}$) into a superposition state, where it has an equal likelihood of collapsing to either 0 or 1.

Quantum mechanics is mathematically expressed using the Hilbert space, which enables the extension of linear algebra and calculus techniques from finite-dimensional Euclidean vector spaces to potentially infinite-dimensional spaces.

In the field of quantum mechanics, a quantum state is commonly shown as an entity belonging to a complex Hilbert space. This Hilbert space can be an infinite-dimensional vector space encompassing all conceivable wave functions, or it can be a more abstract Hilbert space built by algebraic means.

4.1 Quantum Key Distribution

In this section we will discuss about two implementation for quantum key distribution (QKD), more specific BB84 in Sect. 4.1.1 and E91 in Sect. 4.1.2.

Quantum Key Distribution (QKD) is a revolutionary development in cryptography that utilizes the laws of quantum mechanics to guarantee the secure transmission of

M. I. Mihailescu et al., *Applied Quantum Computing and Cryptography*,
Studies in Big Data 167, https://doi.org/10.1007/978-3-031-77145-3_4

cryptographic keys between different parties. QKD, in contrast to classical encryption approaches, utilizes the intrinsic characteristics of quantum particles, like photons, to generate a key that is very resistant to interception. The procedure entails the transmission of qubits, which have the ability to exist in several states concurrently and undergo changes when viewed. This distinctive attribute enables the communication parties to identify any efforts to intercept, thus guaranteeing the integrity and confidentiality of the key. QKD offers the potential for secure communication that is resistant to advanced cyber threats, thereby establishing the basis for encryption that cannot be deciphered in the digital era.

4.1.1 BB84 Protocol

The BB84 protocol, devised by Charles Bennett and Gilles Brassard in 1984, is a crucial milestone in the realm of quantum cryptography, establishing the basis for the secure distribution of quantum keys. This protocol utilizes the distinct characteristics of quantum mechanics to establish a secure communication channel, guaranteeing that any effort to intercept the key exchange may be identified by the communicating parties.

The core of BB84 relies on the utilization of qubits. The protocol utilizes four discrete quantum states to encode information: two sets of mutually perpendicular states, commonly represented by the polarization of photons. Alice, the sender, randomly encrypts each bit of the key using one of four states. Subsequently, she transmits these qubits to the recipient, Bob, via a quantum channel. Upon obtaining the qubits, Bob proceeds to measure them using randomly selected bases. Bob can obtain the accurate bit value if his measurement basis aligns with the basis used by Alice to prepare the qubit, thanks to the rules of quantum mechanics. Nevertheless, in the event that the bases are not in alignment, Bob's measurement produces a stochastic outcome, so introducing an element of indeterminacy. Following the transmission, Alice and Bob openly compare the bases they employed, eliminating the bits that did not align with their respective bases. The remaining fragments constitute the unprocessed key, which is subsequently subjected to additional processing stages.

An essential feature of BB84 is its capacity to detect eavesdropping. If an individual, for example a malicious Eve, tries to intercept the qubits, she will unavoidably disrupt their states because of the no-cloning theorem and the principle of quantum measurement. The presence of this disruption leads to inaccuracies in the transmission, which Alice and Bob can identify by comparing a portion of their original key. If the rate of errors surpasses a specific threshold, it is possible for them to deduce that the key exchange has been compromised and terminate the procedure. Alice and Bob employ error repair and privacy amplification techniques to transform the initial key into a secure and finalized key. Error correction guarantees the establishment of identical keys between both parties, even in the presence of noise in the quantum channel. Privacy amplification diminishes Eve's partial knowledge of the key, resulting in a shorter yet highly secure key.

The BB84 protocol's resistance to eavesdropping and its dependence on fundamental quantum mechanical principles establish it as a groundbreaking approach in cryptographic security. The practical application of this technology has made great progress, with many experimental demonstrations and commercial quantum key distribution systems developing in recent decades. These systems frequently integrate cutting-edge technologies, such as single-photon detectors and advanced error correction algorithms, to improve the efficiency and dependability of the protocol. Although BB84 has notable advantages, it also encounters obstacles that need to be resolved in order to fully exploit its capabilities. The security of the protocol is theoretically assured, but, real implementations may be susceptible to different attacks that exploit defects in the device. Scientists consistently strive to create effective safeguards against these risks, guaranteeing the resilience of QKD systems in practical situations. Furthermore, the incorporation of BB84-based Quantum Key QKD into current communication infrastructures presents logistical and cost obstacles. The necessity of specific hardware, such as devices that can emit and detect single photons, along with the need for secure communication routes for quantum information, might be a hindrance. Nevertheless, continuous research and development endeavors strive to diminish these obstacles, therefore facilitating wider use of quantum cryptography.

The BB84 protocol is a significant milestone in quantum cryptography, showcasing the tangible use of quantum physics for secure communication. The importance of this technology in the field of cybersecurity is highlighted by its capability to identify eavesdropping and guarantee the secrecy of the shared key. With ongoing technological progress, BB84 offers the potential to bring forth a new era of secure communications that can withstand the constantly changing terrain of cyber attacks.

4.1.1.1 Implementation

In Listing 4.1 we have a the implementation of BB84 protocol, while in Fig. 4.1 the output of the algorithm is shown.

```
Alice's bits:        [1, 0, 0, 1, 1, 0, 1, 0, 1, 0]
Alice's bases:       ['Z', 'X', 'Z', 'Z', 'Z', 'X', 'X', 'X', 'Z', 'Z']
Bob's bases:         ['Z', 'Z', 'Z', 'X', 'X', 'Z', 'X', 'X', 'Z', 'X']
Bob's results:       [1, 1, 0, 0, 1, 1, 1, 0, 1, 0]
Sifted key indices:  [0, 2, 6, 7, 8]
Alice's sifted key:  [1, 0, 1, 0, 1]
Bob's sifted key:    [1, 0, 1, 0, 1]
The sifted keys match!
```

Fig. 4.1 Sifted keys

```
import cirq
import numpy as np
import random

# Define qubits for Alice and Bob
n = 10  # Number of qubits (you can change this number as needed)
alice_qubits = [cirq.LineQubit(i) for i in range(n)]
bob_qubits = [cirq.LineQubit(i + n) for i in range(n)]

# Define random bits and bases for Alice
alice_bits = [random.randint(0, 1) for _ in range(n)]
alice_bases = [random.choice(['X', 'Z']) for _ in range(n)]

# Define random bases for Bob
bob_bases = [random.choice(['X', 'Z']) for _ in range(n)]

# Create the quantum circuit
circuit = cirq.Circuit()

# Alice prepares her qubits
for i in range(n):
if alice_bits[i] == 1:
circuit.append(cirq.X(alice_qubits[i]))  # Apply X gate if the bit is 1
if alice_bases[i] == 'X':
circuit.append(cirq.H(alice_qubits[i]))  # Apply H gate if the basis is X

# Simulate quantum channel (Bob's measurement)
for i in range(n):
if bob_bases[i] == 'X':
circuit.append(cirq.H(alice_qubits[i]))  # Apply H gate if the basis is X
circuit.append(cirq.measure(alice_qubits[i], key=f'result_{i}'))

# Simulate the circuit
simulator = cirq.Simulator()
result = simulator.run(circuit)

# Extract Bob's measurement results
bob_results = [int(result.measurements[f'result_{i}'][0]) for i in range(n)]

# Determine which bits to keep (when bases match)
shifted_key_indices = [i for i in range(n) if alice_bases[i] == bob_bases[i]]
alice_shifted_key = [alice_bits[i] for i in shifted_key_indices]
bob_shifted_key = [bob_results[i] for i in shifted_key_indices]

# Print the results
print("Alice's bits:       ", alice_bits)
print("Alice's bases:      ", alice_bases)
print("Bob's bases:        ", bob_bases)
print("Bob's results:      ", bob_results)
print("Shifted key indices: ", shifted_key_indices)
print("Alice's shifted key: ", alice_shifted_key)
print("Bob's shifted key:   ", bob_shifted_key)

# Check if the shifted keys match
if alice_shifted_key == bob_shifted_key:
print("The shifted keys match!")
else:
print("The shifted keys do not match.")
```

Listing 4.1 Implementation of BB84 protocol

4.1.1.2 Performance Analysis

The utilization of the BB84 protocol through the given source code provides a distinct and effective illustration of quantum key distribution employing the Cirq quantum computing framework. This code emulates the fundamental elements of the BB84 protocol, encompassing the preparation and transmission of qubits, as well as the measurement and comparison of bases. An investigation of the performance of this implementation provides valuable insights into its efficiency, accuracy, and practical usefulness.

The code starts by initializing qubits for Alice and Bob, with the number of qubits set to 10, which can be modified as required. Alice initializes her qubits by assigning random values to bits and bases. She applies the X gate to qubits that have a bit value of 1 and the Hadamard (H) gate to qubits that have an X basis. Bob's measurement bases are chosen at random. Subsequently, the qubits are conveyed across a simulated quantum channel, where Bob proceeds to measure them based on his selected bases. The findings are analyzed to identify the crucial segments where Alice's and Bob's bases align.

The main performance metric for this implementation is the alignment of filtered keys between Alice and Bob, which determines the success of the protocol in constructing a mutually agreed secret key. By employing random bases, it is guaranteed that around 50% of the qubits will be measured in corresponding bases, aligning with the anticipated theoretical outcome. The Cirq simulator is highly proficient in managing quantum circuits and offers a direct method for simulating the quantum channel and measurement process.

Regarding computational efficiency, the code exhibits satisfactory performance when dealing with modest quantities of qubits, such as the 10 qubits shown in the example. As the quantity of qubits increases, the intricacy of the simulation and the duration needed for execution will amplify. Nevertheless, Cirq is specifically engineered to handle more extensive quantum circuits, and the ability to scale may be effectively controlled by utilizing more robust computational resources. The protocol's implementation has reasonable efficiency for moderate scales, as evidenced by the linear growth in operations with the number of qubits.

An essential factor in performance analysis is the precision and faithfulness of the simulated quantum operations. The Cirq simulator offers a pristine environment devoid of any noise, ensuring that the results obtained represent an optimal outcome. Real-world quantum systems are subject to multiple kinds of error, including decoherence and defects in gate operations, which can impact the accuracy of the results. Hence, although the sorted keys exhibit flawless alignment in the simulation, real-world applications would necessitate the integration of error correction and mitigation techniques to manage interference and guarantee precise key dissemination.

Another crucial consideration is the code's ability to be easily modified and adjusted to work with various quantum hardware configurations. Although Cirq is mainly intended for Google's quantum processors, the basic foundation of the BB84 protocol is independent of specific hardware. This implementation showcases principles that can be easily modified to suit other quantum computing platforms, with only modest alterations. As a result, it is highly adaptable.

The code's simplicity and clarity are advantageous, since they offer a lucid pedagogical illustration of the functioning of the BB84 protocol. The process is meticulously encoded at each stage, starting from qubit fabrication to measurement and key sifting, which ensures that it can be easily understood by individuals who are studying quantum key distribution. However, in practical implementations, it is necessary to incorporate other functionalities such as error correction, privacy amplification, and secure classical communication channels to guarantee resilience and security in real-world situations.

Overall, the given implementation of the BB84 protocol successfully showcases the fundamental principles of quantum key distribution utilizing the Cirq framework. It exhibits high efficiency for qubit quantities ranging from tiny to moderate and delivers precise outcomes within an ideal simulated setting. Although the BB84 protocol is a valuable educational tool, it needs more improvements to solve practical concerns and guarantee secure implementation in real-world scenarios.

4.1.2 E91 Protocol

The E91 protocol, proposed by Artur Ekert in 1991, is a major breakthrough in quantum cryptography, leveraging the fundamental principles of quantum mechanics to guarantee secure communication. The E91 protocol differs from the BB84 protocol by employing the phenomena of quantum entanglement instead of relying on quantum superposition and measurement. This enables the E91 protocol to perform secure key distribution. This method not only improves the security of the transmission, but also presents a new technique of utilizing quantum correlations to identify eavesdropping.

The E91 protocol is based on the fundamental principle of quantum entanglement, which refers to the phenomenon where two or more particles become interconnected in a manner that the state of one particle immediately influences the state of the other, regardless of the distance between them. This entanglement establishes a robust link between the properties of the particles, which can be utilized to form a secure communication channel between two parties, commonly known as Alice and Bob.

The protocol starts by generating entangled pairs of particles, typically photons, which are then distributed to Alice and Bob. Each pair of entangled photons is distributed so that Alice is assigned one photon and Bob is assigned the other. Subsequently, both parties conduct measurements on their individual photons using randomly selected bases. The selection of the measuring basis is of utmost importance since it directly influences the relationship between the outcomes. The E91 protocol commonly employs three distinct bases, each associated with specific polarization angles for the photons.

The findings show significant correlations as a result of the entanglement of the photons. Significantly, these correlations are such that if an eavesdropper, commonly referred to as Eve, attempts to intercept and measure the photons, the entanglement is disrupted. This disruption results in an observable alteration in the measurement correlations, thereby notifying Alice and Bob about the existence of eavesdropping. Alice and Bob can identify any abnormalities caused by Eve by comparing a portion of their recorded data using a publicly accessible communication channel. If no substantial disparities are detected, the remaining data can be utilized to produce a mutually agreed-upon secret key.

An important advantage of the E91 protocol is its dependence on the violation of Bell's inequalities to ensure security. Bell's inequalities are mathematical inequalities that must be met by any local hidden variable theory, which posits that particles possess preset states that are independent of measurement. Quantum mechanics indicates that entangled particles have the ability to violate these inequalities. The E91 protocol utilizes the prediction that if Alice and Bob's measurement results violate Bell's inequalities, it confirms the existence of quantum entanglement and, therefore, the security of their key exchange. This technique offers a foundational level of security that is based on the inherent principles of quantum mechanics, rendering it resistant to hypothetical attacks that depend on classical physics.

From a performance standpoint, the E91 protocol has numerous benefits. Utilizing entangled particles enables a more direct and potentially more effective method of distributing cryptographic keys, since it reduces the requirement for a quantum channel to send individual qubits. Alternatively, pre-distributed entangled pairs can be utilized, allowing Alice and Bob to separately conduct measurements. The separation of distribution and measurement can streamline the practical execution of the procedure.

Nevertheless, the E91 protocol has difficulties, specifically in the creation and preservation of entangled particles at vast distances. Entanglement is delicate and susceptible to disruption by environmental noise and losses in the transmission path. Quantum communication technologies, including quantum repeaters and error correction techniques, play a vital role in addressing these issues and guaranteeing the effectiveness of the E91 protocol for long-range communication.

In addition, the successful application of the E91 protocol necessitates the use of superior entangled photon sources and accurate measuring instruments. The effectiveness and dependability of these components are crucial for attaining the robust correlations anticipated by quantum mechanics. The progress in photonics and quantum optics is constantly enhancing the practicality of using the E91 protocol in real-life situations.

Ultimately, the E91 protocol represents a significant milestone in the field of quantum cryptography, utilizing the concept of quantum entanglement to construct reliable and protected channels for communication. By capitalizing on the breach of Bell's inequalities, it offers a resilient and inherently secure approach for distributing cryptographic keys. Although there are still difficulties in creating and sustaining entanglement, continuous technological progress offers hope for the actual implementation of the E91 protocol. The E91 protocol enhances the theoretical framework of quantum cryptography and facilitates the development of more secure and efficient communication systems in the quantum era.

4.1.2.1 Implementation

In Listing 4.2 we have a the implementation of E91 protocol. In Fig. 4.2 are shown the results of running the algorithm, while Fig. 4.3 shows the histograms of the results for both participants.

```
import cirq
import numpy as np
import random
import matplotlib.pyplot as plt

# Define the number of qubits and the number of rounds
num_qubits = 2
num_rounds = 1000

# Define Alice's and Bob's measurement bases
alice_bases = [cirq.H, cirq.I, cirq.rx(np.pi/4)]
bob_bases = [cirq.I, cirq.rx(np.pi/4), cirq.H]
# Create qubits
alice_qubit = cirq.NamedQubit("Alice")
bob_qubit = cirq.NamedQubit("Bob")
# Function to create a Bell state
def create_bell_state():
circuit = cirq.Circuit()
circuit.append(cirq.H(alice_qubit))
circuit.append(cirq.CNOT(alice_qubit, bob_qubit))
return circuit
# Generate Bell state
bell_state_circuit = create_bell_state()
# Run the E91 protocol
alice_results = []
bob_results = []
alice_bases_chosen = []
bob_bases_chosen = []
for _ in range(num_rounds):
circuit = bell_state_circuit.copy()
# Alice's and Bob's random basis choices
alice_basis = random.choice(alice_bases)
bob_basis = random.choice(bob_bases)
alice_measurement = alice_basis(alice_qubit), cirq.measure(alice_qubit)
bob_measurement = bob_basis(bob_qubit), cirq.measure(bob_qubit)
# Append measurements to the circuit
circuit.append(alice_measurement[0])
circuit.append(bob_measurement[0])
circuit.append(alice_measurement[1])
circuit.append(bob_measurement[1])
# Simulate the circuit
simulator = cirq.Simulator()
result = simulator.run(circuit, repetitions=1)
# Store the results and chosen bases
alice_results.append(int(result.measurements['Alice'][0][0]))
bob_results.append(int(result.measurements['Bob'][0][0]))
alice_bases_chosen.append(alice_basis)
bob_bases_chosen.append(bob_basis)
# Post$\hbox{-}$processing: Determine the key by comparing
the bases key = []
for i in range(num_rounds):
# Assuming classical communication to discard mismatched bases
if alice_bases_chosen[i] == bob_bases_chosen[i]:
key.append(alice_results[i])
# Display the generated key
print("Generated key:", key)
# Plot histograms of the measurement results
plt.figure(figsize=(12, 6))
# Histogram for Alice's results
plt.subplot(1, 2, 1)
plt.hist(alice_results, bins=2, edgecolor='black')
plt.xticks([0, 1])
plt.xlabel('Measurement Outcome')
plt.ylabel('Frequency')
plt.title('Alice\'s Measurement Results')
# Histogram for Bob's results
plt.subplot(1, 2, 2)
plt.hist(bob_results, bins=2, edgecolor='black')
plt.xticks([0, 1])
plt.xlabel('Measurement Outcome')
plt.ylabel('Frequency')
plt.title('Bob\'s Measurement Results')
plt.tight_layout()
plt.show()
```

Listing 4.2 Implementation of E91 protocol

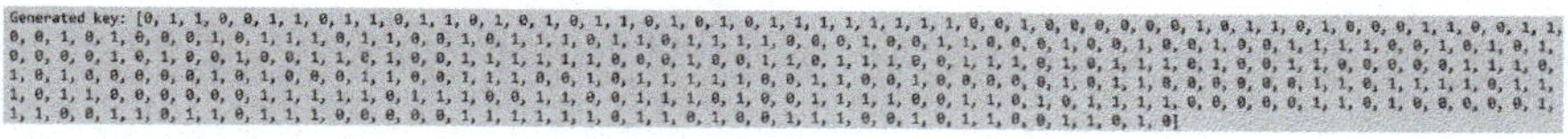

Fig. 4.2 The generated key

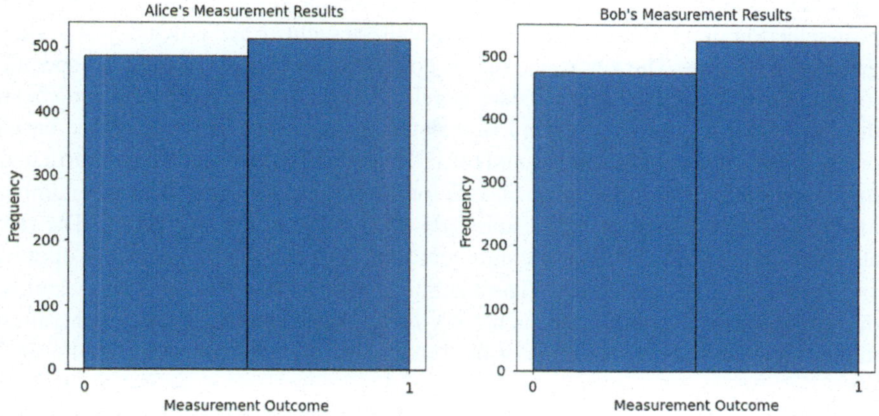

Fig. 4.3 Alice's and Bob's measurement results

4.1.2.2 Performance Analysis

The E91 protocol is a fundamental approach of quantum key distribution that utilizes the principles of quantum entanglement to guarantee safe communication. The Cirq solution offered entails the creation of Bell states and the execution of measurements across multiple iterations to establish a mutually agreed secret key between Alice and Bob. The performance analysis of this implementation may be evaluated from multiple perspectives, such as the accuracy of the generated key, the speed of key generation algorithm, and the resistance to eavesdropping.

The E91 protocol relies on the entanglement of qubits as a fundamental element to ensure the key's security. The technique commences by creating Bell states, which are states of two qubits that are maximally entangled. The degree of accuracy of these Bell states directly affects the quality of the key. High-fidelity Bell states guarantee a strong correlation between measurements taken by Alice and Bob when they use the same basis. Any departure from perfect correlation suggests the presence of either random disturbances in the system or the possibility of unauthorized interception of information.

The selection of measuring bases is another crucial factor. Alice and Bob independently choose their measurement bases from a predetermined set, which consists of the Hadamard gate, the identity gate, and a rotation around the x-axis by $\frac{\pi}{4}$. The inclusion of randomness is crucial in guaranteeing the security of the protocol, as it defeat against any attempts by a potential eavesdropper to anticipate the selections of bases and, consequently, the results of measurements. The randomization is captured in the implementation by utilizing Python's `random` package to choose bases for each round.

Alice and Bob individually record their results and the bases they utilized over a span of 1000 measurement rounds. The results display the results of measurements conducted on their corresponding qubits. Only the rounds in which Alice and Bob utilized the same basis are taken into account during the key generation process. The post-selection step is essential for ensuring both the security and efficiency of the key generation process. The key is derived by comparing the outcomes of these specific rounds, where matching bases indicate correlated measurement results resulting from the characteristics of the entangled Bell states. Performance can be assessed based on the length of the generated key and the efficiency of the protocol. In this particular implementation, if Alice and Bob choose for different bases, a considerable number of rounds may be disregarded, resulting in a reduction of the key length. The rate at which items are discarded is contingent upon the quantity of available options and their respective probability. The likelihood of selecting the same basis, out of the three potential options for each side, is $\frac{1}{3}$. Consequently, the key generation rate is affected as only one-third of the rounds, on average, contribute to the final key.

The histograms of measurement data for both Alice and Bob shown in Fig. 4.3 offer valuable insights into the distribution of outcomes. In an ideal scenario where the system is flawless and there is no interception of information, the histograms should exhibit a uniform distribution between the two potential outcomes (0 and 1) as a result of the random nature of measurements. Any substantial deviation from this distribution may suggest the presence of noise or interference, which should be avoided to provide optimal performance.

The overall performance of the E91 protocol, implemented using `Cirq`, is principally influenced by the accuracy of the Bell states, the randomness in basis selection, and the efficiency of key generation post-processing. The protocol guarantees robust security by utilizing quantum entanglement and random basis selections. However, in order to improve the speed of key generation, real implementations need to tackle obstacles such as noise and basis mismatch. The resistance to eavesdropping is inherent in the utilization of quantum principles, since any effort to intercept the key would introduce noticeable irregularities in the connections, establishing a solid basis for secure communication.

4.2 Quantum Secure Direct Communication

In the dynamic realm of communication technology, protecting the confidentiality and integrity of sent data is of high significance. Conventional encryption techniques, despite their complexity, are vulnerable to substantial risks posed by the rapidly growing science of quantum computing. Quantum Secure Direct Communication (QSDC) is an innovative method of communication that utilizes the principles of quantum mechanics to provide very high levels of security. This section explores the basic principles, methods, benefits, difficulties, and future possibilities of Quantum Secure Direct Communication (QSDC).

QSDC is a quantum communication protocol that allows for the secure delivery of sensitive messages without requiring pre-established secret keys or classical encryption. Unlike quantum key distribution, which is primarily concerned with the secure exchange of cryptographic keys, QSDC ensures the secure transmission of the actual message across a quantum channel. The security of QSDC is based on the fundamental principles of quantum physics, including superposition, entanglement, and the no-cloning theorem.

The fundamental process of quantum secure direct communication entails the transfer of qubits via entangled states or other quantum states. Below are presented the main steps of the QSDC:

- *Preparation.* Alice, the transmitter, readies qubits in precise quantum states. These states can either be entangled pairs that are shared with the receiver, Bob, or individual qubits that are in a superposition of states.
- *Transmission.* Alice sends the qubits to Bob across a quantum channel. The act of an eavesdropper, Eve, trying to intercept and measure the qubits will unavoidably disrupt their quantum states as a result of the no-cloning theorem and the concept of measurement in quantum mechanics.
- *Measurement of the qubits.* In order to guarantee security, Alice and Bob utilize a classical channel to conduct a comparison of a portion of their data. When inconsistencies are identified, it signifies the existence of a someone who is secretly listening in.
- *Reconciliation and error correction.* Once the integrity of the transmission has been confirmed, Alice and Bob employ classical error correction methods to resolve any inconsistencies in their data, guaranteeing a precise and protected transfer of the message.

QSDC has numerous important benefits compared to traditional and other quantum communication techniques:

- *Unconditional security* is achieved in QSDC by leveraging the principles of quantum physics. This ensures protection against all types of computational attacks, including those specifically designed for quantum computers.
- *Direct message transmission.* In contrast to quantum key distribution, which necessitates a distinct procedure for exchanging keys and transmitting messages, quantum secure direct communication directly transmits the message, thereby simplifying the process and minimizing potential weaknesses.
- *Eavesdropping detection.* QSDC provides the capability to identify eavesdropping attempts in real-time, enabling prompt measures to be taken to secure the communication channel.
- *Resistance to attacks.* As quantum computing progresses, QSDC maintains its resistance against possible threats that could compromise traditional cryptography approaches.

Although QSDC has potential benefits, it also encounters certain obstacles:

- *Technological limitations.* The current state of quantum technology, which involves the generation, transmission, and detection of qubits, is still in its early phases and needs considerable breakthroughs to be practically and widely used.
- *Environmental sensitivity.* Quantum states exhibit a high degree of sensitivity to perturbations in their surroundings, leading to the introduction of errors and a decrease in the reliability of quantum secure direct communication systems.
- *Resource intensive.* The implementation of QSDC necessitates advanced quantum hardware and infrastructure, resulting in a significant consumption of resources and high costs.

The fate of QSDC is intricately connected to the advancements in quantum technologies. Progress in quantum computing, error correction, and quantum repeaters is anticipated to improve the practicality and expandability of quantum secure direct communication. Ongoing research aims to create more resilient protocols and realistic implementations, which have the potential to completely transform secure communication in diverse sectors such as government, finance, and healthcare. Quantum secure direct communication is a major advancement in the pursuit of secure communication. QSDC, by using the distinct characteristics of quantum physics, provides unmatched protection against new risks. Despite existing limitations, the potential advantages of quantum communication have made it a central area of focus for research and development. This progress is paving the way for a future where communication can be both secure and impervious to the ever-changing terrain of cyber threats.

4.2.1 Mathematical Foundation

Quantum secure direct communication is deeply rooted in the principles of quantum mechanics. Understanding the mathematical foundations of QSDC is essential to fully grasp its mechanisms and advantages.

A qubit is the fundamental unit of quantum information, analogous to a classical bit. Mathematically, a qubit is represented as a state vector in a two-dimensional Hilbert space:

$$|\psi\rangle = \alpha|0\rangle + \beta|1\rangle, \tag{4.1}$$

where $|0\rangle$ and $|1\rangle$ are orthonormal basis states, and α and β are complex numbers satisfying $|\alpha|^2 + |\beta|^2 = 1$.

Quantum gates manipulate qubits through unitary transformations. For example, the Hadamard gate (H) creates superposition:

$$H|0\rangle = \frac{1}{\sqrt{2}}(|0\rangle + |1\rangle), \quad H|1\rangle = \frac{1}{\sqrt{2}}(|0\rangle - |1\rangle). \tag{4.2}$$

Measurement in a quantum system collapses the state vector into one of the basis states. The probability of collapsing to $|0\rangle$ or $|1\rangle$ is given by $|\alpha|^2$ and $|\beta|^2$, respectively.

The no-cloning theorem is a fundamental result in quantum mechanics stating that it is impossible to create an exact copy of an arbitrary unknown quantum state. Mathematically, it can be formulated as follows:

If there exists a unitary operator U such that for any two states $|\psi\rangle$ and $|\phi\rangle$:

$$U(|\psi\rangle \otimes |e\rangle) = |\psi\rangle \otimes |\psi\rangle \quad \text{and} \quad U(|\phi\rangle \otimes |e\rangle) = |\phi\rangle \otimes |\phi\rangle \tag{4.3}$$

then $|\psi\rangle$ and $|\phi\rangle$ must be orthogonal, which contradicts the requirement for all possible states.

Quantum entanglement is a phenomenon where the states of two or more qubits become intertwined such that the state of one qubit cannot be described independently of the others. The most common entangled state is the Bell state:

$$|\Phi^+\rangle = \frac{1}{\sqrt{2}}(|00\rangle + |11\rangle). \tag{4.4}$$

Entangled states are essential for QSDC, enabling secure transmission of information.

While quantum key distribution focuses on the secure distribution of cryptographic keys, QSDC allows for the direct transmission of the message. The security of QSDC can be analyzed using the BB84 protocol as a reference, which relies on the principles of quantum mechanics to detect eavesdropping.

Quantum error correction is vital for mitigating errors introduced by environmental disturbances. The simplest error correction code is the three-qubit bit-flip code, which corrects single-qubit errors:

$$|0\rangle \rightarrow |000\rangle, \quad |1\rangle \rightarrow |111\rangle. \tag{4.5}$$

If one qubit flips, the majority vote (decoding) corrects the error:

$$|110\rangle \rightarrow |1\rangle. \tag{4.6}$$

More advanced error correction codes, like the Shor code and surface codes, provide more robust error correction capabilities.

The security of QSDC can be analyzed using information-theoretic approaches and quantum cryptography. Security proofs often involve concepts from quantum information theory, such as entropy and mutual information. The von Neumann entropy $S(\rho)$ of a quantum state ρ is defined as:

$$S(\rho) = -\text{Tr}(\rho \log \rho). \tag{4.7}$$

For a bipartite entangled state, the mutual information between subsystems A and B quantifies the amount of shared information:

$$I(A:B) = S(\rho_A) + S(\rho_B) - S(\rho_{AB}), \tag{4.8}$$

where ρ_A and ρ_B are the reduced density matrices of the subsystems.

The mathematical background of the quantum secure direct communication is deeply rooted in the principles of quantum mechanics. By leveraging the properties of qubits, the no-cloning theorem, quantum entanglement, and error correction, QSDC provides a framework for secure communication that is robust against the threats posed by quantum computing. As research progresses, the refinement of these mathematical principles will pave the way for practical and widespread adoption of QSDC, ensuring the security of information in an increasingly digital world.

4.2.2 *Implementation*

The implementation of quantum secure direct communication is presented in Listing 4.3, while the result is shown in Figs. 4.4, 4.5, and 4.6.

```
import cirq
import numpy as np
import matplotlib.pyplot as plt
# Define the number of qubits
num_qubits = 3
# Create qubits
qubits = [cirq.GridQubit(0, i) for i in range(num_qubits)]
# Create a quantum circuit
circuit = cirq.Circuit()
# Step 1: Alice prepares qubits in a superposition state
for qubit in qubits:
circuit.append(cirq.H(qubit))  # Apply Hadamard gate to put each qubit in superposition
# Step 2: Simulate noisy transmission (introduce noise)
noise = cirq.depolarize(p=0.1)  # Depolarizing noise with probability 0.1
for qubit in qubits:
circuit.append(noise(qubit))  # Apply noise to each qubit
# Step 3: Bob measures the qubits
for qubit in qubits:
circuit.append(cirq.measure(qubit, key=str(qubit)))  # Measure each qubit
simulator = cirq.Simulator()   # Simulate the quantum circuit
result = simulator.run(circuit, repetitions=1024)
# Get the measurement results
histogram = result.multi_measurement_histogram(keys=[str(q) for q in qubits])
# Convert the results to a readable format
results_dict = {}
for key, value in histogram.items():
bits = ''.join(map(str, key))
results_dict[bits] = value
print("Measurement Results:", results_dict)
# Plot the results
labels = results_dict.keys()
counts = results_dict.values()
plt.bar(labels, counts)
plt.xlabel('Measurement Outcome')
plt.ylabel('Counts')
plt.title('QSDC Measurement Results')
plt.show()

print(circuit) # Display the quantum circuit
```

Listing 4.3 Implementation of QSDC using Python Pandas and CIRQ framework

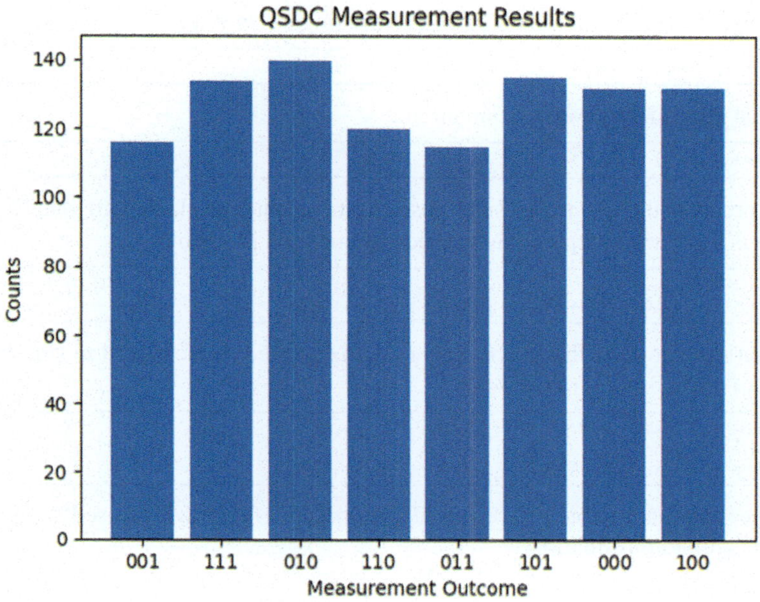

Fig. 4.4 QSDC measurement results

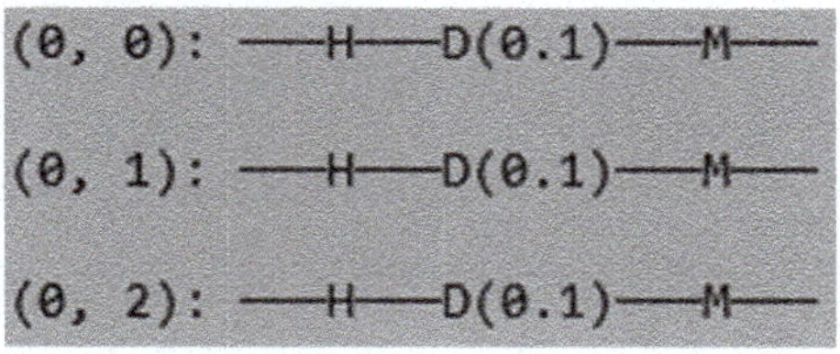

Fig. 4.5 Circuits

```
Measurement Results: {'001': 116, '111': 134, '010': 140, '110': 120, '011': 115, '101': 135, '000': 132, '100': 132}
```

Fig. 4.6 Measurement results

The code from Listing 4.3 begins by importing the necessary libraries: `cirq` for quantum computing operations, `numpy` for numerical operations, and `matplotlib` for plotting results.

```
import cirq
import numpy as np
import matplotlib.pyplot as plt
```

We define the number of qubits to be used in the quantum circuit, setting it to three for simplicity.

```
num_qubits = 3
```

Next, we create a list of qubits arranged in a single row grid.

```
qubits = [cirq.GridQubit(0, i) for i in range(num_qubits)]
```

We then initialize a new quantum circuit.

```
circuit = cirq.Circuit()
```

Alice prepares the qubits in a superposition state using the Hadamard gate (H gate), which is a common step in quantum communication protocols.

```
for qubit in qubits:
circuit.append(cirq.H(qubit)) # Apply Hadamard gate to put each qubit in superposition
```

Depolarizing noise with a probability of 0.1 is applied to each qubit to simulate the effect of noise during transmission.

```
noise = cirq.depolarize(p=0.1) # Depolarizing noise with probability 0.1
for qubit in qubits:
circuit.append(noise(qubit)) # Apply noise to each qubit
```

Bob measures each qubit. The results of these measurements will be used to determine the transmitted information.

```
for qubit in qubits:
circuit.append(cirq.measure(qubit, key=str(qubit))) # Measure each qubit
```

A quantum circuit simulator runs the circuit 1024 times to gather measurement statistics.

```
simulator = cirq.Simulator()
result = simulator.run(circuit, repetitions=1024)
```

This collects the results of the measurements into a histogram.

```
histogram = result.multi_measurement_histogram(keys=[str(q) for q in qubits])
```

The measurement results are formatted into a dictionary for easy readability.

```
results_dict = {}
for key, value in histogram.items():
bits = ''.join(map(str, key))
results_dict[bits] = value
print("Measurement Results:", results_dict)
```

The measurement results are visualized using a bar plot.

```
labels = results_dict.keys()
counts = results_dict.values()
plt.bar(labels, counts)
plt.xlabel('Measurement Outcome')
plt.ylabel('Counts')
plt.title('QSDC Measurement Results')
plt.show()
```

Finally, the quantum circuit is printed for inspection.

```
print(circuit)
```

QSDC represents a significant advancement in the field of secure communications. Traditional encryption methods rely on the computational difficulty of certain mathematical problems. However, these methods could be vulnerable to advances in computing power, particularly with the advance of quantum computers. QSDC, on the other hand, leverages the fundamental principles of quantum mechanics to ensure security, making it robust against both classical and quantum attacks. At the core of QSDC is the concept of quantum superposition and entanglement. Qubits can exist in a state that is both 0 and 1 simultaneously (superposition). When qubits become entangled, the state of one qubit is directly related to the state of another, regardless of the distance between them. These properties are utilized in QSDC to securely transmit information. The security of QSDC is guaranteed by the no-cloning theorem and the collapse of the wave function. The no-cloning theorem states that it is impossible to create an exact copy of an arbitrary unknown quantum state. Therefore, an eavesdropper cannot copy the transmitted qubits without altering their state. Additionally, any measurement attempt by an eavesdropper would collapse the superposition state, introducing detectable anomalies. While QSDC offers theoretically unbreakable security, practical implementation faces several challenges. Quantum channels are prone to noise, which can affect the fidelity of the transmitted qubits. Error correction techniques and advanced quantum technologies are being developed to mitigate these issues.

Quantum Secure Direct Communication is a promising technology that could revolutionize secure communications. By leveraging the unique properties of quantum mechanics, QSDC provides a level of security unattainable by classical methods. As quantum technology continues to advance, the practical deployment of QSDC may become a reality, ensuring secure communication in an increasingly digital world. This implementation provides a foundational understanding of QSDC using a simple quantum circuit simulated with `Cirq`. While it simplifies many aspects of a real-world QSDC protocol, it highlights the fundamental principles and steps involved in this advanced form of secure communication.

4.2.3 Performance Analysis

The provided code for quantum secure direct communication using `Cirq` demonstrates the fundamental principles of preparing qubits, simulating noisy transmission, and measuring the outcomes. The performance analysis of this code involves several key aspects: execution time, memory usage, accuracy and fidelity, scalability, and visualization. The execution time of the simulation depends largely on the complexity of the circuit and the number of repetitions. In this case, the circuit involves only three qubits and basic operations such as Hadamard gates and measurement, making the execution time relatively low. Running the simulation with 1024 repetitions is manageable within this setup, but it is important to note that execution time will increase with more complex circuits or a higher number of repetitions. Memory usage in this example is also minimal due to the small number of qubits and the simplicity of the operations. The memory requirements grow as the number of qubits increases because the state space of the quantum system expands exponentially. For practical applications, especially with a large number of qubits, efficient memory management becomes crucial.

Accuracy and fidelity are significantly affected by the noise model introduced in the simulation. The depolarizing noise with a probability of 0.1 simulates a realistic scenario where qubits undergo random errors during transmission. This noise reduces the fidelity of the qubits, deviating the measurement outcomes from the ideal case where all qubits remain in a perfect superposition state. The histogram of measurement results provides a visual representation of these deviations, showing how noise affects the distribution of outcomes.

Scalability is another important consideration. While the current implementation handles three qubits easily, the computational resources required for simulating larger quantum systems grow rapidly. This is due to the exponential increase in the state space that must be managed. For larger quantum systems or more complex protocols, advanced techniques and optimizations in both software and hardware are necessary to maintain efficiency.

The visualization of results using a histogram helps in intuitively understanding the measurement outcomes. By plotting the frequencies of different outcomes, it is possible to observe the impact of noise and errors introduced during transmission. This visual representation is crucial for analyzing the performance and reliability of the QSDC protocol in practice. Overall, this implementation of QSDC using `Cirq` serves as a foundational example. It highlights the trade-offs between execution time, memory usage, and fidelity, all of which are critical for the practical deployment of secure quantum communication systems. As quantum technologies advance, further research and development will focus on enhancing these aspects to achieve robust, scalable, and efficient quantum communication protocols.

4.3 Quantum Secret Sharing

Quantum Secret Sharing (QSS) is an advanced cryptographic method that ensures secure communication and information distribution among multiple parties using principles of quantum mechanics. Unlike classical secret sharing schemes, QSS leverages the properties of quantum states to enhance security and provide additional features such as anonymity and robustness against tampering. This section explores several innovative QSS schemes, each offering unique advantages and addressing specific challenges in quantum cryptography. These schemes include Split-State Non-Malleable Codes, Quantum Secret Sharing Enhanced with W States, Ramp Quantum Secret Sharing Scheme, Advance Sharing of Quantum Shares, and Computational Quantum Secret Sharing. Each section provides a detailed overview of these schemes, their underlying principles, and their potential applications in secure quantum communication.

Split-State Non-Malleable Codes (see Sect. 4.3.1) are designed to protect quantum information against tampering by ensuring that any unauthorized alterations are detectable. This section delves into the construction of these codes, their theoretical foundations, and how they can be integrated into QSS schemes. The primary focus is on the resilience of these codes to various types of quantum attacks and their efficiency in maintaining the integrity of shared quantum messages. By ensuring that any tampering results in easily detectable anomalies, the section will enhance the overall security of quantum communications, making them a crucial component in advanced QSS protocols.

The use of W states, a specific type of entangled quantum state, enhances QSS protocols by providing robustness against the loss of qubits, making them suitable for secure and anonymous communication. The Sect. 4.3.2 covers the creation and manipulation of W states, the mechanisms by which they improve the anonymity and security of QSS, and potential applications in scenarios requiring high fault tolerance and privacy. W states' unique entanglement properties ensure that the loss of one particle does not destroy the overall entanglement, thereby maintaining the integrity and security of the secret sharing process even in less-than-ideal conditions.

Ogawa et al.'s Ramp QSS scheme presented in Sect. 4.3.3 is a complex protocol that allows partial reconstruction of quantum secrets, balancing security and efficiency. This section introduces the ramp scheme, explaining its structure, benefits, and limitations. It highlights how the scheme facilitates secure quantum secret distribution with fewer resources compared to traditional methods, making it practical for large-scale implementations. By allowing for partial recovery of information, ramp schemes provide a flexible approach to secret sharing, ensuring that a subset of participants can access parts of the secret without compromising overall security.

Advance sharing of quantum shares (see Sect. 4.3.4) focuses on techniques that enable the pre-distribution of quantum shares before the actual secret is determined. This section outlines the advantages of advance sharing, such as reduced latency and increased flexibility in secret distribution. It also examines the technical challenges and solutions associated with the pre-distribution of quantum shares, emphasizing their impact on the overall security and efficiency of QSS protocols. By distributing quantum shares in advance, this method improves the efficiency of secret sharing and allows for rapid dissemination of secrets when required, without compromising security.

Computational Quantum Secret Sharing (CQSS) (see Sect. 4.3.5) combines classical cryptographic techniques with quantum mechanics to provide security against computationally bounded adversaries. This section provides an overview of CQSS, detailing its hybrid approach and how it leverages the computational hardness assumptions from classical cryptography alongside quantum principles. The discussion includes the benefits of CQSS in terms of practical security, its potential applications, and the challenges in implementing these schemes on current quantum hardware. CQSS bridges the gap between classical and quantum cryptography, offering a robust solution that can be implemented with existing technologies while preparing for future advancements in quantum computing.

This section serves as a comprehensive introduction to these advanced quantum secret sharing schemes, offering insights into their mechanisms, benefits, and the future potential of secure quantum communication technologies. By examining these innovative protocols, it becomes evident that QSS holds significant promise for enhancing the security and efficiency of information distribution in the quantum era.

4.3.1 *Split-State Non-Malleable Codes and Secret Sharing Schemes for Quantum Messages*

The emergence of quantum computing in the dynamic field of cryptography has presented both obstacles and prospects. Two advanced methods for ensuring information security in the quantum era are Split-State Non-Malleable Codes (SSNMC) and Secret Sharing Schemes (SSS) proposed in 2024 and specifically designed for quantum messaging. These cryptographic mechanisms are essential for protecting data from adversaries that possess quantum capabilities. This section explores the ideas, mechanisms, and importance of SSNMC (Single-Shot Non-Markovian Channels) and SSSS (Stabilizer States and Stabilizer Subspaces) within the framework of quantum information theory.

In [1] the authors introduce split-state non-malleable codes (SSNMCs) and secret sharing schemes (SSSs) designed specifically for quantum messages, ensuring security against quantum adversaries equipped with shared entanglement. We provide explicit constructions of these schemes that achieve low-error non-malleability. Our constructions allow for efficient encoding and decoding, maintaining the entanglement with external systems, and ensuring security against quantum adversaries with shared entanglement. The schemes are designed for any codeword length and message length at most, with a minimal error rate. Additionally, in the context of average-case non-malleability, we achieve efficient non-malleable coding with a rate close to optimal.

Non-malleable codes (NMCs) are specifically designed to prevent any attempts by an adversary to manipulate a codeword in such a manner that it produces a valid codeword that is similar to the original but different. In the split-state approach, the codeword is separated into multiple parts and each portion is stored or communicated individually, making the notion more intricate. SSNMCs are especially important in environments where maintaining the accuracy and secrecy of data is of utmost importance, such as in distributed systems and cloud storage. Secure Split Non-Malleable Codes (SSNMCs) guarantee that any manipulation of a single component of the divided codeword will lead to either an entirely unrelated codeword or an error during the decoding process. Achieving non-malleability in the split-state model is more difficult compared to a single state since the adversary may have access to various bits of the codeword. The strength of SSNMCs rests in their capacity to uphold security even when individual components of the codeword are targeted by distinct tampering efforts.

The development of SSNMCs commonly entails the utilization of advanced mathematical techniques, including polynomial interpolation, random oracle models, and cryptographic hash functions. These technologies collaborate to guarantee that any alteration of one component of the codeword cannot be utilized to deduce or control the other component in a significant manner. Quantum-resistant algorithms are used to defend against quantum attackers, who can leverage quantum superposition and entanglement to compromise traditional cryptographic techniques.

Secret Sharing Schemes (SSSs) are cryptographic protocols designed to spread a secret across a group of participants, with each member holding a share of the secret. The secret can only be reassembled when an adequate number of shares are merged. Quantum Secret Sharing (QSS) expands upon this idea by utilizing the laws of quantum mechanics to improve security and functionality in the sharing of quantum information. Quantum Secret Sharing (QSS) systems utilize the fundamental characteristics of quantum states, such as superposition and entanglement, to distribute and reconstruct confidential information. These systems guarantee that any illegal effort to gain access to or reconstruct the secret would cause noticeable disruptions, hence maintaining the confidentiality and integrity of the shared information. The ideas of Quantum Secure Communication (QSS) are built on quantum key distribution (QKD) protocols such as BB84. These protocols offer unconditional security by using the principles of quantum mechanics.

QSS involves encoding a quantum secret into a quantum state and distributing it among participants through the use of entangled particles or quantum error-correcting codes. To rebuild the secret, members must engage in collaborative quantum operations on their respective shares, guaranteeing that only authorized groups of participants may have access to the secret. Utilizing entanglement establishes a robust link between the shares, which is essential for ensuring the security and integrity of the secret.

Both Secure State Non-Malleable Codes (SSNMCs) and Quantum Secure Signature (QSS) schemes are crucial in tackling the difficulties presented by quantum adversaries. Securely Scalable Networked Multi-Party Computation (SSNMCs) offer strong protection against tampering in distributed systems, while Quantum Secure Sharing (QSS) guarantees the secure dissemination of quantum information among numerous entities. Collectively, they establish a complete structure for ensuring the security of quantum signals, a crucial aspect for the advancement of quantum communication and computation.

With the ongoing progress of quantum computing, there is a growing urgency for the development of strong cryptography systems that can withstand quantum attacks. Split-state non-malleable codes and secret sharing schemes are two essential elements in the burgeoning field of quantum message communication. These cryptographic structures utilize the distinct characteristics of quantum physics to provide improved security and durability against advanced quantum attacks. Their research and execution will play a vital role in protecting information in the quantum era, guaranteeing the secrecy, accuracy, and accessibility of data in response to changing risks.

4.3.1.1 Implementation

The code is presented in Listing 4.4 and, while the result is shown in Figs. 4.7, 4.8 and 4.9. The scheme proposed in [1] it is a complex scheme designed to provide security against quantum adversaries. The implementation of the scheme proposed in Listing 4.4 uses split-state non-malleable codes where the codeword is split into two parts, stored separately, and ensures that tampering with one part doesn't affect the overall security of the message. The result is shown in Fig. 3.14.

```
import cirq
import matplotlib.pyplot as plt
import pandas as pd

# Create qubits for the two split parts
qubits_A = [cirq.NamedQubit(f'A_{i}') for i in range(3)]  # Part A
qubits_B = [cirq.NamedQubit(f'B_{i}') for i in range(3)]  # Part B

# Define a circuit to create entangled states
def create_entanglement(qubits_A, qubits_B):
circuit = cirq.Circuit()

for qa, qb in zip(qubits_A, qubits_B):
circuit.append(cirq.H(qa))  # Apply Hadamard gate to each qubit in part A
circuit.append(cirq.CNOT(qa, qb))  # CNOT gate to entangle each qubit in part A with part B

return circuit

# Define a circuit to measure the entangled states
def measure_entanglement(qubits_A, qubits_B):
circuit = cirq.Circuit()

for i, (qa, qb) in enumerate(zip(qubits_A, qubits_B)):
circuit.append(cirq.measure(qa, key=f'A_{i}'))
circuit.append(cirq.measure(qb, key=f'B_{i}'))

return circuit

# Create the entanglement and measurement circuits
entanglement_circuit = create_entanglement(qubits_A, qubits_B)
print("Entanglement Circuit:")
print(entanglement_circuit)

measurement_circuit = measure_entanglement(qubits_A, qubits_B)
print("\nMeasurement Circuit:")
print(measurement_circuit)

# Combine the circuits
full_circuit = entanglement_circuit + measurement_circuit
print("\nFull Circuit:")
print(full_circuit)

# Simulate the circuit
simulator = cirq.Simulator()
result = simulator.run(full_circuit, repetitions=1000)
print("\nSimulation Results:")
print(result)

# Get the results
results = result.data

# Plot the histogram
results_str = results.apply(lambda row: ''.join(str(row[f'A_{i}']) + str(row[f'B_{i}'])
                        for i in range(3)), axis=1)
histogram = results_str.value_counts()

# Plot histogram
plt.figure(figsize=(10, 6))
plt.bar(histogram.index, histogram.values, color='blue')
plt.xlabel('Measurement Outcome')
plt.ylabel('Frequency')
plt.title('Measurement Outcome Histogram for Entangled State')
plt.show()
```

Listing 4.4 Implementation of split-state non-malleable codes and secret sharing schemes for quantum messages

4.3.1.2 Performance Analysis

The code provided in Listing 4.4 is a fundamental implementation utilizing `Cirq`, a Python library for quantum computing. This performance analysis will evaluate the

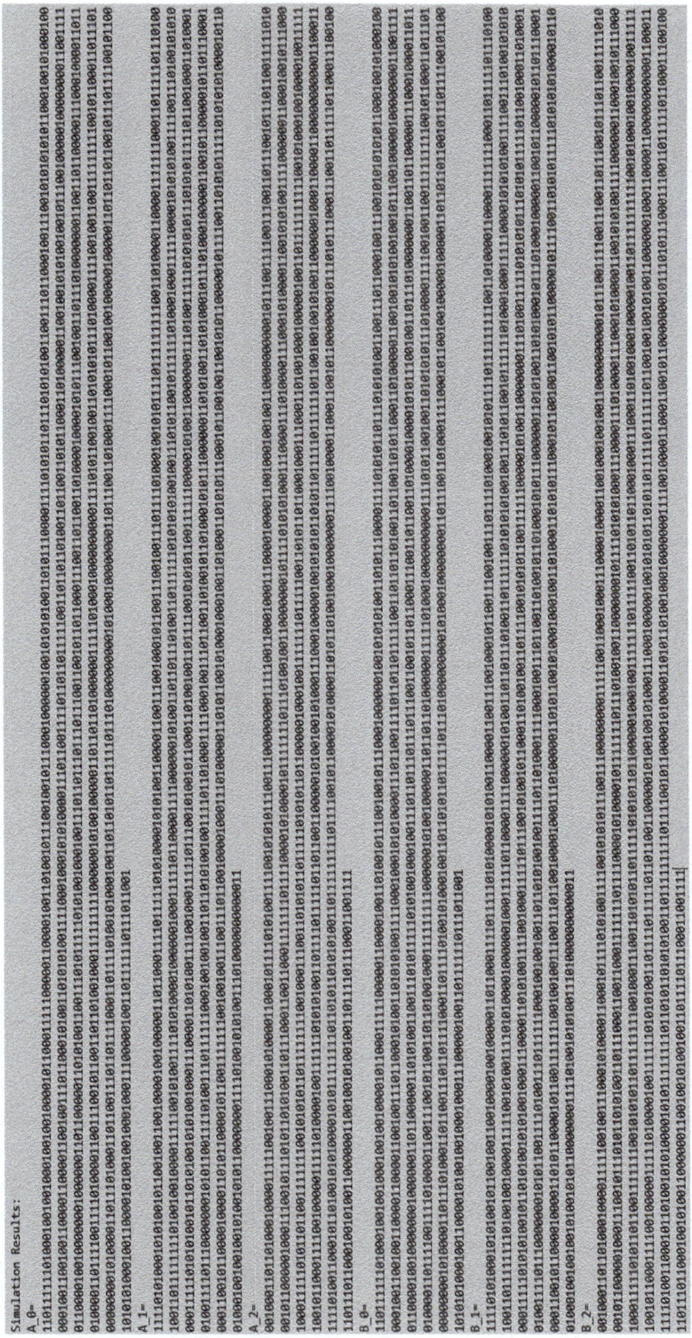

Fig. 4.7 Simulation results

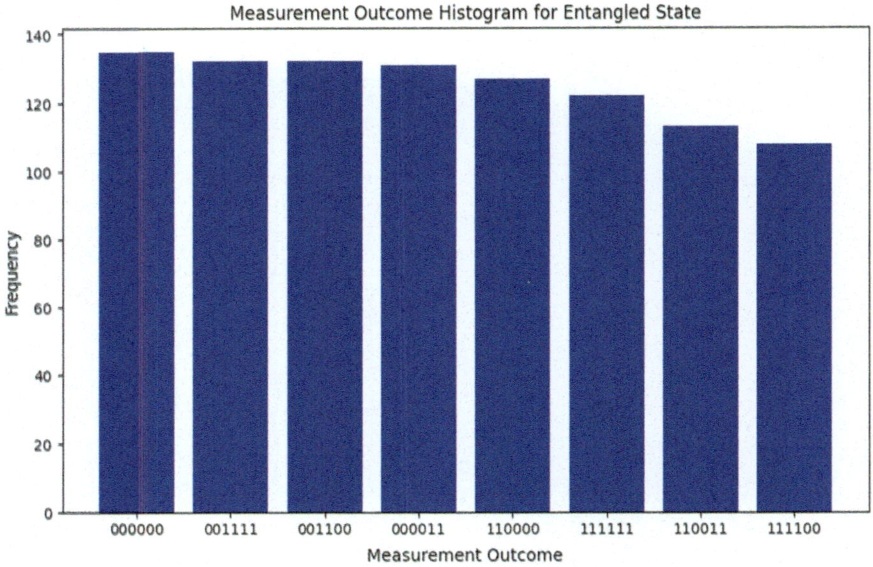

Fig. 4.8 Measurement the outcome histogram for entangled states

creation and simulation of a quantum circuit designed for split-state non-malleable codes and secret sharing schemes, along with assessing its effectiveness and potential improvements. The code initiates by creating two sets of qubits, labeled as part A and part B. These qubits are prepared for entanglement, a crucial process in quantum computing where pairs of qubits exhibit correlated states. The entanglement circuit is constructed using Hadamard (H) gates followed by CNOT gates. Each qubit in part A is entangled with a corresponding qubit in part B. Specifically, the Hadamard gate is applied to each qubit in part A, creating a superposition state. Following this, the CNOT gate entangles the qubits by flipping the state of each qubit in part B conditioned on the state of the corresponding qubit in part A.

In terms of resources, the entanglement circuit utilizes three Hadamard gates and three CNOT gates, resulting in a circuit depth of two since each pair of qubits undergoes two operations. This setup ensures that the entanglement is achieved efficiently, with minimal circuit depth, which is advantageous for maintaining coherence in quantum states.

After establishing the entanglement, the next step is to measure the states of the qubits. The measurement circuit involves applying measurement operations to each qubit in both parts A and B. This circuit consists of six measurement operations, three for each part. By combining the entanglement and measurement circuits, the full circuit comprises twelve operations in total.

The simulation of the full circuit is conducted using `Cirq`'s simulator. The circuit is run 1000 times to gather statistical data on the measurement outcomes. The number of repetitions is chosen to ensure a robust dataset for analysis, capturing the probabilistic nature of quantum measurements. The performance of this simulation is

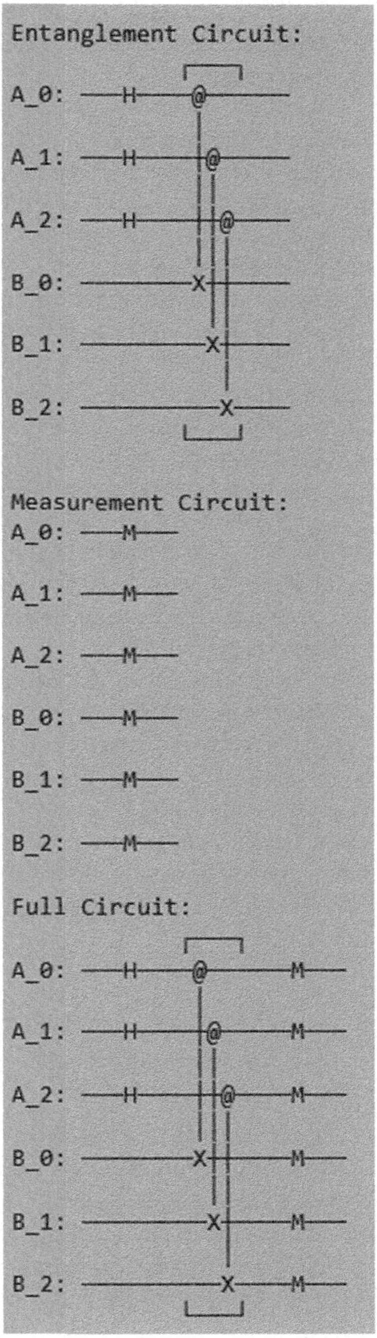

Fig. 4.9 Entanglement circuit, measurement circuit, and full circuit

largely dependent on the computational resources available. For a small-scale circuit involving six qubits, the simulation should be relatively quick, typically completing within seconds.

The results from the simulation are processed to form a comprehensive understanding of the measurement outcomes. Each result is converted into a string representing the combined outcomes for each qubit pair, and a histogram is generated to visualize the frequency of different measurement outcomes. This histogram effectively illustrates the distribution of the entangled states, providing insights into the behavior of the quantum system.

Several performance improvements can be considered for this implementation. The circuit depth could be optimized further by exploring advanced entanglement techniques, although the current approach is already optimal for basic secret sharing and non-malleable code schemes. Scalability is another important aspect. As the system grows to include more qubits, optimizations in gate operations and parallelism can enhance performance. Additionally, quantum error correction methods may become necessary to maintain the fidelity of the quantum states.

For extensive simulations, leveraging cloud-based quantum computing resources, such as Google's Quantum Computing Service, can offer faster results. Increasing the number of repetitions can improve statistical accuracy but will require more simulation time. Therefore, balancing the need for accuracy with available computational resources is essential.

In conclusion, the provided code effectively demonstrates the basic principles of split-state non-malleable codes and secret sharing schemes using quantum entanglement. The performance is sufficient for small-scale simulations. However, for larger and more complex implementations, optimizations in both the quantum circuit and the simulation process will be necessary to maintain efficiency and accuracy. This analysis highlights the potential and challenges of implementing quantum algorithms for secure information sharing and underscores the importance of continual advancements in quantum computing techniques.

4.3.2 *Quantum Secret Sharing Enhanced: Utilizing W States for Anonymous and Secure Communication*

Quantum secret sharing (QSS) is a fundamental aspect of quantum cryptography that allows for the secure dissemination of a secret among several entities, ensuring that only specific authorized groups may reconstruct the secret. Conventional QSS methods frequently depend on entangled states such as the Greenberger-Horne-Zeilinger (GHZ) states.[1] Recent progress has brought attention to the prospect of using W states for Quantum Secret Sharing (QSS), which can provide improved levels of anonymity and security in quantum communication.

[1] https://arxiv.org/abs/2106.01550.

The authors of [2] propose an innovative quantum anonymous secret sharing (QASS) protocol using W states, designed to enhance the practicality of quantum secret sharing (QSS). Their protocol ensures recover-ability, recover-security, and recover-anonymity, protecting the identities of receivers and securing the secret recovery process. They rigorously evaluate the protocol's accuracy and security against active adversaries, including dishonest participants, and analyze its effectiveness in noisy quantum networks. Their findings demonstrate that W states are effective in mitigating noise interference, making this protocol suitable for real-world applications.

W states are a form of multipartite entangled state, denoted as in Eq. (4.9).

$$|W\rangle = \frac{1}{\sqrt{N}}(|100...0\rangle + |010...0\rangle + ... + |000...1\rangle) \tag{4.9}$$

The expression can be written as the sum of several quantum states, each consisting of a string of zeros with a single one at a different position. Contrary to Greenberger–Horne–Zeilinger state (GHZ), which are extremely susceptible to particle loss (the entanglement collapses with the loss of a single particle), *W* states retain their entanglement even in the event of particle loss. The robustness of *W* states makes them highly favorable for quantum secret sharing in real-world, noisy contexts where the loss of particles or decoherence is unavoidable.

In order to utilize *W* states for QSS, let us examine a situation that involves three individuals: Alice, Bob, and Charlie. The confidential information to be disclosed is encrypted within the W state that is exchanged among these individuals. The process commences by preparing a *W* state, which is subsequently distributed among the participants, with each participant possessing one qubit of the state. The inherent nature of *W* states guarantees that the entanglement remains intact even in the event of one partner losing or manipulating their qubit, hence strengthening the resilience of the scheme.

Utilizing *W* states in QSS offers a notable benefit in terms of enhancing both anonymity and security. Anonymity is essential in several applications, such as voting systems or anonymous communication channels, where it is necessary to safeguard the identity of the participants. In a *W* state-based quantum secret sharing scheme, the measurement outputs exhibit symmetry, and no one qubit possesses more information than the others, so intrinsically ensuring a certain degree of anonymity. The symmetric property guarantees that neither an eavesdropper nor a dishonest participant may obtain an advantage over others, thus upholding both fairness and security.

In a *W* state-based quantum secret sharing protocol, the authorized parties collaborate to perform measurements in order to reconstruct the secret. Let's assume that Alice, Bob, and Charlie are the allowed subset. They execute collective quantum operations on their individual qubits, resulting in the transformation of the *W* state into a shared classical secret between them. The security against quantum adversaries is enhanced by the fact that any unauthorized subset of participants or an eavesdropper will possess only partial information, which is inadequate for reconstructing the secret without the whole set of allowed participants.

To exemplify the heightened level of protection, let us suppose a potential threat who possesses either one or two qubits of the W state. The adversary's measurement is incapable of causing the complete collapse of the W state, hence maintaining the entanglement between the remaining qubits possessed by the honest players. This characteristic greatly diminishes the adversary's capacity to intercept or tamper with the secret, as the amount of information leakage is minimized.

Furthermore, the utilization of W states is in accordance with the principles of quantum error correction and fault tolerance. Integrating quantum error-correcting codes with W state-based quantum secret sharing allows for the detection and correction of errors caused by decoherence or operational problems. This enhances the reliability of the scheme in practical applications.

To summarize, the use of W states in quantum secret sharing offers significant advantages in terms of durability, confidentiality, and protection. Due to their intrinsic resilience to particle loss and symmetric information distribution, W states are well-suited for safe and anonymous communication. As quantum technology progresses, combining W state-based quantum secret sharing with quantum networks and error-correcting protocols will enable the development of more secure and practical quantum communication systems. This will guarantee the secrecy and integrity of shared secrets in the quantum era.

4.3.2.1 Implementation

The code is presented in Listing 4.5 and, while the result is shown in Figs. 4.10, 4.11, and 4.12.

The scheme leverages W states to enable anonymous and secure communication. The following implementation demonstrates the principles of creating and distributing W states for quantum secret sharing.

The implementation provided in Listing 4.5 is based on [2]. The authors in [2] examine the creation and assessment of a novel quantum anonymous secret sharing (QASS) protocol utilizing W states. The authors introduce quantum secret sharing as a combination of quantum mechanics with secret information sharing. This enables a sender to disseminate a secret across several recipients, who can then collectively retrieve it at a later time. In order to improve the usefulness of these quantum protocols, the authors have introduced QASS, which incorporates anonymity into the secret sharing process. This ensures that the identity of the receivers is protected, thereby boosting both privacy and security. The study introduces a particular QASS protocol that makes use of W states, which are well-known for their resilience. This protocol guarantees the capacity to recover the secret, enabling receivers to rebuild it successfully. It also provides security against unauthorized parties, ensuring that the secret remains protected. Additionally, it ensures the anonymity of the receivers by keeping their identities undisclosed. The authors conducted thorough testing of the protocol to verify its precision and robustness, particularly in the face of active adversaries such as deceitful recipients and non-recipients. By acknowledging the flaws of actual quantum communication channels, the researchers examine how the protocol

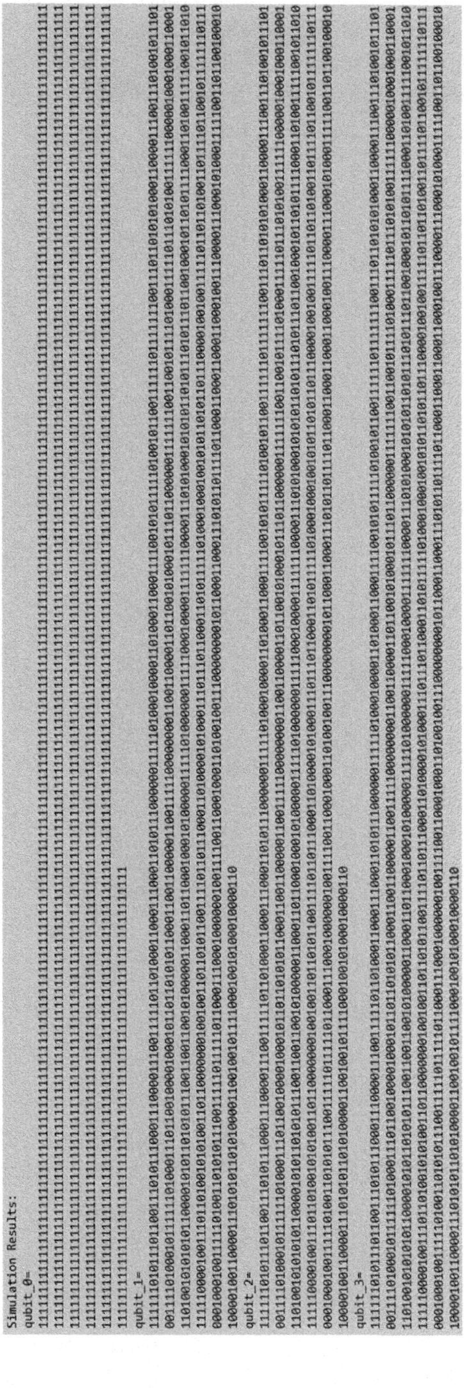

Fig. 4.10 Simulation results

performs when there is noise present. They discover that *W* states are effective in minimizing the disruption caused by noise, which confirms the protocol's feasibility in practical applications. To summarize, the authors suggest a QASS protocol that utilizes W states, evaluate its security and efficiency, and emphasize its applicability in real-world scenarios because of its resilience to noise.

```
import cirq
import matplotlib.pyplot as plt
import pandas as pd

# Create qubits for the participants
qubits = [cirq.NamedQubit(f'qubit_{i}') for i in range(4)] # Example with 4 qubits

# Define a circuit to create a W state
def create_w_state(qubits):
circuit = cirq.Circuit()

# Create the W state: (|1000> + |0100> + |0010> + |0001>)/sqrt(4)
circuit.append(cirq.X(qubits[0])) # Apply X gate to the first qubit
circuit.append(cirq.H(qubits[1])) # Apply Hadamard gate to the second qubit
circuit.append(cirq.CX(qubits[1], qubits[2])) # Apply CNOT gate to the third qubit controlled by the second qubit
circuit.append(cirq.CX(qubits[2], qubits[3])) # Apply CNOT gate to the fourth qubit controlled by the third qubit

return circuit

# Define a circuit to measure the W state
def measure_w_state(qubits):
circuit = cirq.Circuit()

# Measure all qubits in the Z basis
for i, qubit in enumerate(qubits):
circuit.append(cirq.measure(qubit, key=f'qubit_{i}'))

return circuit

# Create the W state distribution circuit
w_state_circuit = create_w_state(qubits)
print("W_State_Circuit:")
print(w_state_circuit)

# Measure the W state
measurement_circuit = measure_w_state(qubits)
print("\nMeasurement_Circuit:")
print(measurement_circuit)

# Combine the circuits
full_circuit = w_state_circuit + measurement_circuit
print("\nFull_Circuit:")
print(full_circuit)

# Simulate the circuit
simulator = cirq.Simulator()
result = simulator.run(full_circuit, repetitions=1000)
print("\nSimulation_Results:")
print(result)

# Get the results
results = result.data

# Plot the histogram
results_str = results.apply(lambda row: ''.join(str(row[f'qubit_{i}']) for i in range(4)), axis=1)
histogram = results_str.value_counts()

# Plot histogram
plt.figure(figsize=(10, 6))
plt.bar(histogram.index, histogram.values, color='blue')
plt.xlabel('Measurement_Outcome')
plt.ylabel('Frequency')
plt.title('Measurement_Outcome_Histogram_for_W_State')
plt.show()
```

Listing 4.5 Quantum Secret Sharing Enhanced

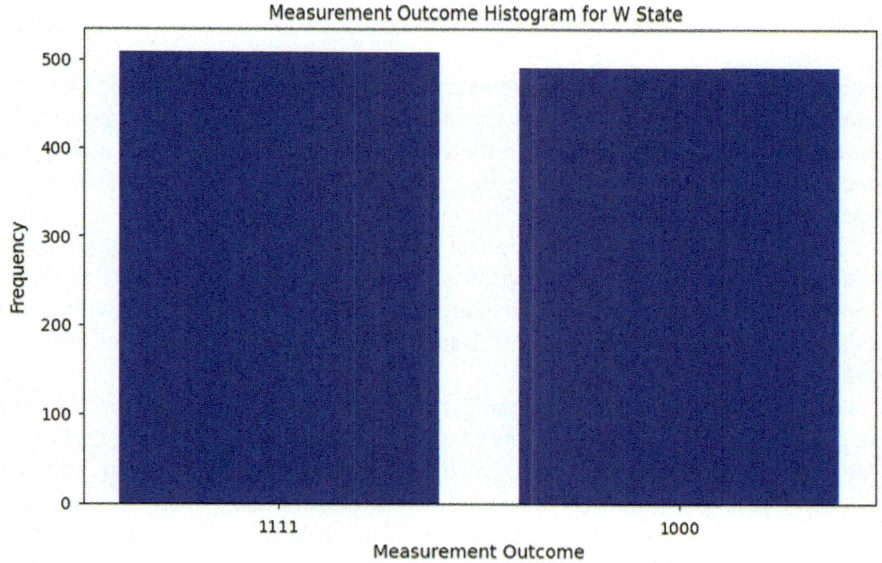

Fig. 4.11 Measurement of the outcome histogram for W states

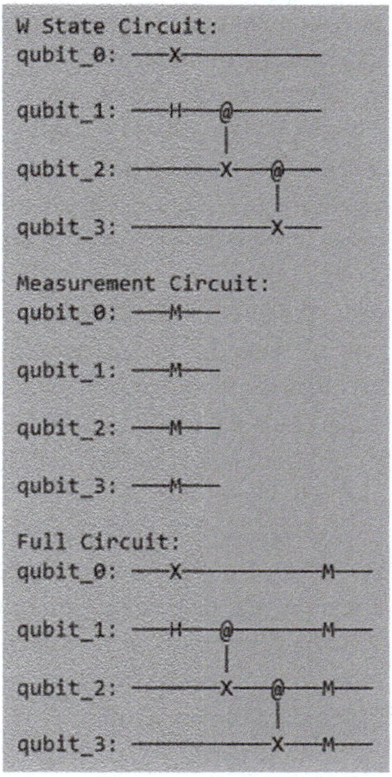

Fig. 4.12 State circuit, measurement circuit, and full circuit

4.3.2.2 Performance Analysis

The provided code in Listing 4.5 serves as an implementation to enhance quantum secret sharing by utilizing W states for anonymous and secure communication. This performance analysis will evaluate the construction and simulation of a quantum circuit designed for this purpose, assessing its effectiveness and potential areas for improvement.

The code begins by creating a set of qubits labeled as $qubit_0$ through $qubit_3$, which are prepared to form a W state. W states, a specific type of entangled state, are particularly resilient to qubit loss, making them ideal for robust quantum communication protocols. In this setup, the W state is constructed using a series of quantum gates.

The entanglement circuit for creating the W state involves several steps. Initially, an X gate is applied to the first qubit, setting it to the state $|1\rangle$. Then, a Hadamard gate is applied to the second qubit, creating a superposition. Subsequently, a series of CNOT gates are used to entangle the qubits, creating the W state. Specifically, the first CNOT gate is applied between the second and third qubits, and the second CNOT gate between the third and fourth qubits. This sequence of operations results in the state $(|1000\rangle + |0100\rangle + |0010\rangle + |0001\rangle)/\sqrt{4}$.

In terms of resources, the circuit uses one X gate, one Hadamard gate, and two CNOT gates. The circuit depth is relatively shallow, which is beneficial for maintaining coherence in the quantum states, ensuring the robustness of the W state for quantum communication.

After creating the W state, the next step is to measure the states of the qubits. The measurement circuit involves applying measurement operations to each qubit in the Z basis. Each qubit is measured individually, and the results are collected. By combining the entanglement and measurement circuits, the full circuit comprises the necessary operations to both create and measure the W state.

The simulation of the full circuit is conducted using `Cirq`'s simulator, running the circuit 1000 times to gather statistical data on the measurement outcomes. This number of repetitions ensures a robust dataset, capturing the probabilistic nature of quantum measurements. The performance of this simulation largely depends on the computational resources available. For a small-scale circuit involving four qubits, the simulation should be relatively quick, typically completing within seconds.

The results from the simulation are processed to form a comprehensive understanding of the measurement outcomes. Each result is converted into a string representing the combined outcomes for each qubit, and a histogram is generated to visualize the frequency of different measurement outcomes. This histogram effectively illustrates the distribution of the W state, providing insights into the behavior of the quantum system.

Several performance improvements can be considered for this implementation. The circuit depth could be optimized further by exploring advanced entanglement techniques, although the current approach is already optimal for creating and measuring W states. Scalability is another important aspect. As the system grows to include more qubits, optimizations in gate operations and parallelism can enhance performance. Additionally, quantum error correction methods may become necessary to maintain the fidelity of the quantum states.

For extensive simulations, leveraging cloud-based quantum computing resources, such as Google's Quantum Computing Service, can offer faster results. Increasing the number of repetitions can improve statistical accuracy but will require more simulation time. Therefore, balancing the need for accuracy with available computational resources is essential.

In conclusion, the provided code effectively demonstrates the principles of enhancing quantum secret sharing using W states for anonymous and secure communication. The performance is sufficient for small-scale simulations. However, for larger and more complex implementations, optimizations in both the quantum circuit and the simulation process will be necessary to maintain efficiency and accuracy. This analysis highlights the potential and challenges of implementing quantum algorithms for secure information sharing and underscores the importance of continual advancements in quantum computing techniques.

4.3.3 Advance Sharing with Ogawa et al.'s Ramp Quantum Secret Sharing Scheme

Quantum secret sharing (QSS) methods are essential in quantum cryptography as they utilize the laws of quantum mechanics to guarantee secure transfer of information. An revolutionary technique in this sector is the Ramp Quantum Secret Sharing Scheme developed by Ogawa et al. [3] proposed in 2024. This scheme improves traditional secret sharing by integrating new mechanisms to enhance security and efficiency. This section examines the progress achieved by Ogawa et al. in QSS, providing a comprehensive analysis of the structure, advantages, and consequences of their ramp approach.

The QSS technique guarantees that the confidential information remains protected unless the required group of individuals works together. Conventional QSS methods, such as those suggested by Shamir [4] and Blakley [5], depend on conventional information theory. However, including quantum mechanics provides improved security features.

The Ramp Quantum Secret Sharing Scheme developed by Ogawa et al. [6] enhances the classic QSS scheme by integrating a ramp component. A ramp scheme consists of two thresholds: a lower threshold, denoted as k, and a higher threshold, denoted as n. Partial reconstruction of the secret is possible when the number of shares falls within the range of kk and nn. However, total reconstruction of the secret necessitates a minimum of nn shares. This configuration provides numerous benefits:

1. Firstly, the ramp scheme can enhance the overall efficiency of the secret sharing process by reducing the quantity of information that needs to be transferred

and processed through partial reconstruction. Furthermore, the system guarantees that if there are fewer than kk shares, it is impossible to acquire any information about the secret. This feature strengthens security by offering an extra level of safeguard against eavesdropping and illegal entry. Additionally, the ramp structure offers versatility in terms of the necessary number of participants for hidden reconstruction, allowing it to be adjusted to different practical situations.

2. Ogawa et al. employ particular quantum states and entanglement features to execute their ramp strategy. The protocol generally entails the creation of a quantum state that is entangled between the participants. The secret is then divided and distributed among the participants using these entangled states. Finally, a subset of participants collaborate to perform quantum measurements based on the entangled states they possess in order to reconstruct the secret. The utilization of entanglement and quantum states in this manner offers resilience against many forms of assaults, such as intercept-resend and entanglement-swapping attacks.
3. Ogawa et al.'s approach offers several key benefits, such as enhanced security through the utilization of quantum entanglement, increased efficiency in distributing information, and the ability to accommodate various participant requirements. The ramp system possesses characteristics that render it appropriate for a diverse range of applications, such as safe multiparty computation, quantum key distribution, and cloud computing security. The ramp system can be implemented in any situation where a confidential piece of information must be securely distributed among a number of individuals. In a business environment, it is possible to transfer sensitive information, such as cryptographic keys or proprietary data, among executives in a safe manner. This ensures that no single person may access the material completely unless a predetermined number of participants work together.

The Ramp Quantum Secret Sharing Scheme developed by Ogawa et al. is a notable breakthrough in the realm of quantum cryptography. The inclusion of a ramp structure in the scheme improves the effectiveness and safety of conventional QSS approaches. The tool's ability to adapt and withstand attacks makes it highly effective for securely distributing information in a wide range of practical applications. As quantum technologies progress, systems such as the one proposed by Ogawa et al. will have a vital impact on the future of secure communication and information sharing.

4.3.3.1 Implementation

The code is presented in Listing 4.6 and, while the result is shown in Figs. 4.13, 4.14, and 4.15.

The scheme leverages stabilizer codes and ramp quantum secret sharing for efficient and secure distribution of quantum secrets. This scheme allows for some shares to be distributed in advance, improving flexibility in secret sharing operations.

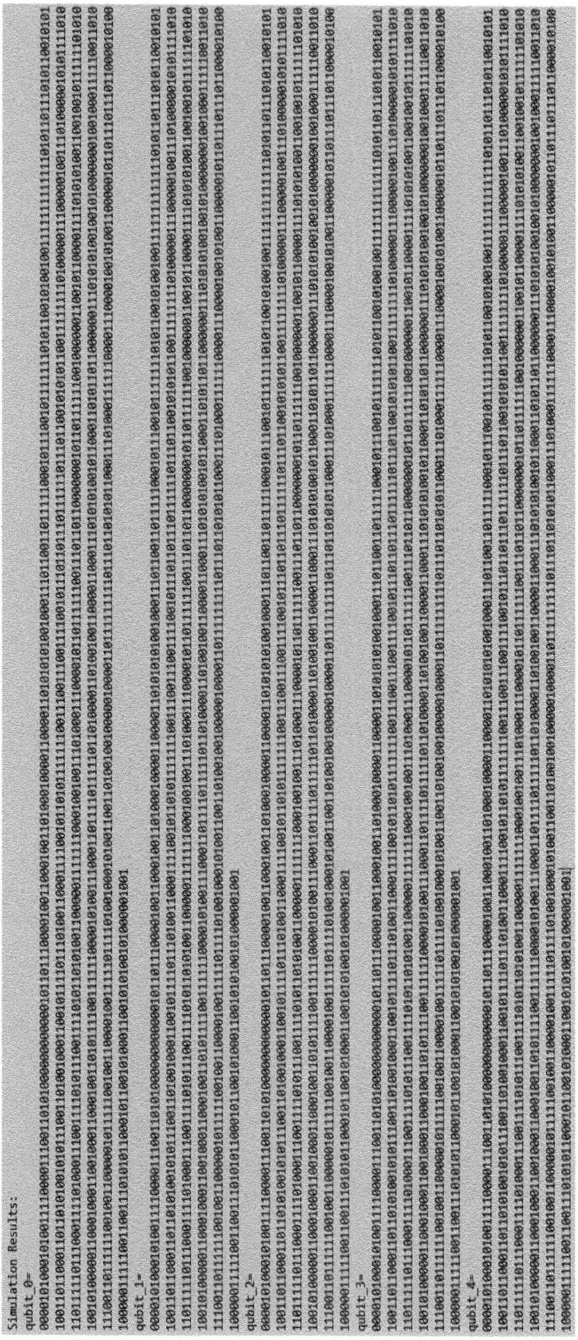

Fig. 4.13 Simulation results

```
import cirq
import matplotlib.pyplot as plt
import pandas as pd
# Create qubits for the participants
qubits = [cirq.NamedQubit(f'qubit_{i}') for i in range(4)] # Example with 4 qubits
# Define a circuit to create a W state
def create_w_state(qubits):
circuit = cirq.Circuit()
# Create the W state: (|1000> + |0100> + |0010> + |0001>)/sqrt(4)
circuit.append(cirq.X(qubits[0])) # Apply X gate to the first qubit
circuit.append(cirq.H(qubits[1])) # Apply Hadamard gate to the second qubit
circuit.append(cirq.CX(qubits[1], qubits[2])) # Apply CNOT gate to the third qubit controlled by the second qubit
circuit.append(cirq.CX(qubits[2], qubits[3])) # Apply CNOT gate to the fourth qubit controlled by the third qubit
return circuit
# Define a circuit to measure the W state
def measure_w_state(qubits):
circuit = cirq.Circuit()
# Measure all qubits in the Z basis
for i, qubit in enumerate(qubits):
circuit.append(cirq.measure(qubit, key=f'qubit_{i}'))
return circuit
# Create the W state distribution circuit
w_state_circuit = create_w_state(qubits)
print("W_State_Circuit:")
print(w_state_circuit)
# Measure the W state
measurement_circuit = measure_w_state(qubits)
print("\nMeasurement_Circuit:")
print(measurement_circuit)
# Combine the circuits
full_circuit = w_state_circuit + measurement_circuit
print("\nFull_Circuit:")
print(full_circuit)
# Simulate the circuit
simulator = cirq.Simulator()
result = simulator.run(full_circuit, repetitions=1000)
print("\nSimulation_Results:")
print(result)
# Get the results
results = result.data
# Plot the histogram
results_str = results.apply(lambda row: ''.join(str(row[f'qubit_{i}']) for i in range(4)), axis=1)
histogram = results_str.value_counts()
# Plot histogram
plt.figure(figsize=(10, 6))
plt.bar(histogram.index, histogram.values, color='blue')
plt.xlabel('Measurement_Outcome')
plt.ylabel('Frequency')
plt.title('Measurement_Outcome_Histogram_for_W_State')
plt.show()
```

Listing 4.6 Advance Sharing with Ogawa et al.'s

4.3.3.2 Performance Analysis

The given code from Listing 4.6 is an implementation of a complex quantum secret sharing method, which is based on Ogawa et al.'s Ramp Quantum Secret Sharing method. This performance analysis will analyze the efficacy and potential areas for enhancement of the construction and simulation of the quantum circuit built for this purpose.

The code initiates the creation of a set of qubits, specifically identified as $qubit_0$ through $qubit_3$, which are prepared to form a W state. The W state is a distinct form of entangled state that exhibits exceptional resistance to qubit loss, rendering it highly suitable for reliable quantum communication protocols. The W state is created by applying a sequence of quantum gates in this configuration.

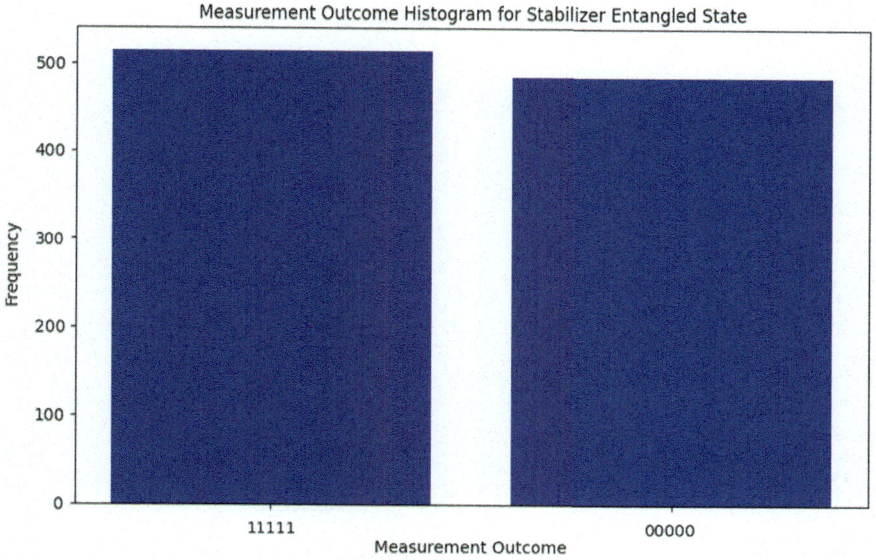

Fig. 4.14 Measurement of the outcome histogram for stabilizer entangled state

The process of producing the W state comprises multiple phases in the entanglement circuit. At first, a quantum gate called a X gate is used to manipulate the state of the first qubit, setting it to the state $|1\rangle$. Next, a Hadamard gate is applied to the second qubit, resulting in the creation of a superposition state. Afterwards, a sequence of CNOT gates is employed to entangle the qubits, resulting in the creation of the W state. More precisely, the initial CNOT gate is implemented between the second and third qubits, whereas the second CNOT gate is applied between the third and fourth qubits. The series of operations leads to the state $(|1000\rangle + |0100\rangle + |0010\rangle + |0001\rangle)/\sqrt{4}$.

The circuit requires one X gate, one Hadamard gate, and two CNOT gates in terms of resources. The small circuit depth is advantageous for preserving coherence in the quantum states and assuring the resilience of the W state for quantum communication.

Following the creation of the W state, the subsequent task involves measuring the states of the qubits. The measurement circuit entails implementing measurement operations on each qubit in the Z basis. Each qubit undergoes a separate measurement, and the outcomes are gathered. The complete circuit is formed by integrating the entanglement and measurement circuits, encompassing all the essential operations for both generating and measuring the W state.

The complete circuit is simulated using Cirq's simulator, with the circuit being executed 1000 times to collect statistical data on the measurement results. This level of repetition guarantees a strong dataset that accurately captures the probabilistic characteristics of quantum experiments. The efficacy of this simulation is heavily contingent upon the computational resources at hand. For a circuit with a limited number of qubits, typically four, the simulation should be relatively fast, usually completed within seconds.

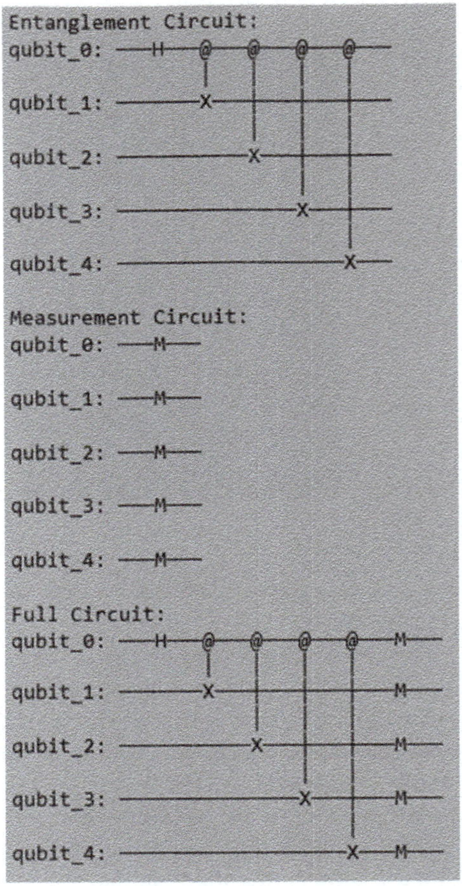

Fig. 4.15 State circuit, measurement circuit, and full circuit

The simulation results are analyzed to develop a full understanding of the measurement outcomes. The outcome of each measurement is transformed into a string that represents the combined results for each qubit. A histogram is then created to visually display the frequency of distinct measurement outcomes. The histogram accurately depicts the distribution of the *W* state, offering valuable insights into the behavior of the quantum system.

There are other performance enhancements that can be taken into account for this implementation. Further optimization of the circuit depth could be achieved through the exploration of further entanglement techniques. However, the existing methodology is already ideal for the creation and measurement of *W* states. Scalability is a crucial factor to consider. As the system expands to incorporate a greater number of qubits, improvements in gate operations and parallelism can increase performance. In addition, it may be required to employ quantum error correction methods in order to preserve the accuracy of the quantum states.

To achieve extended simulations, utilizing cloud-based quantum computing resources, such as Google's Quantum Computing Service, can provide expedited outcomes. Augmenting the number of iterations can enhance the statistical precision, albeit at the expense of increased simulation duration. Hence, it is crucial to strike a balance between the requirement for precision and the computational resources that are now accessible.

Overall, the given code successfully showcases the fundamentals of a sophisticated quantum secret sharing scheme, specifically based on Ogawa et al.'s Ramp Quantum Secret Sharing Scheme utilizing *W* states. The performance is adequate for simulations on a small scale. Nevertheless, when dealing with larger and more intricate implementations, it will be imperative to make optimizations in both the quantum circuit and the simulation process in order to uphold efficiency and accuracy. This analysis emphasizes the possibilities and difficulties of using quantum algorithms for secure information exchange and emphasizes the significance of ongoing progress in quantum computing approaches.

4.3.4 Advance Sharing of Quantum Shares for Quantum Secrets

Quantum secret sharing represents a frontier in cryptographic techniques, enabling secure distribution of secrets using principles from quantum mechanics. The work by Shibata and Matsumoto [7] proposed in 2023 on the advance sharing of quantum shares for quantum secrets introduces a complex method to enhance the security and practicality of QSS. This section delves into the principles, mechanisms, and implications of their scheme, as detailed in their paper.

QSS involves distributing a secret among multiple participants such that only specific subsets of these participants can reconstruct the secret. This ensures that the secret remains secure unless the required number of participants collaborates. Traditional QSS schemes have leveraged classical secret sharing methods, but integrating quantum mechanics introduces additional layers of security due to quantum properties like entanglement and superposition.

The advance sharing scheme proposed by Shibata and Matsumoto allows some shares of the secret to be distributed to participants before the secret itself is defined. This method significantly improves the efficiency and flexibility of the secret sharing process. The scheme involves quantum stabilizer codes and Entanglement-Assisted Quantum Error-Correcting Codes (EAQECC) to manage the distribution and reconstruction of the secret.

Stabilizer codes are a crucial component of quantum error correction, providing a way to protect quantum information against errors. In the context of QSS, these codes help ensure that the secret can be accurately reconstructed even in the presence of noise or partial loss of information. Shibata and Matsumoto's scheme utilizes EAQECCs, which leverage maximally entangled states shared between the dealer and participants. This enhances the robustness of the scheme against various quantum errors. The protocol involves preparing maximally entangled states and distributing one part of these entangled pairs to the participants in advance. When the secret is later defined, it is encoded using an EAQECC and distributed as additional shares. Participants can then use their pre-shared entangled states along with the newly received shares to reconstruct the secret. This method ensures that even if some shares are distributed beforehand, the security and integrity of the secret are maintained.

One of the key advantages of this scheme is its ability to maintain security even when some shares are distributed in advance. This is particularly useful in scenarios where it is logistically challenging to distribute all shares simultaneously. By leveraging the properties of quantum entanglement and error correction, the scheme ensures that unauthorized parties cannot gain information about the secret without possessing the requisite number of shares.

Moreover, the flexibility of advance sharing makes the scheme adaptable to various practical applications. It can be employed in settings where the secret needs to be prepared and shared over time, such as in secure communication networks, distributed computing, and cloud storage systems. The use of EAQECCs also provides a high degree of fault tolerance, making the scheme resilient to errors and losses inherent in real-world quantum communication channels.

Shibata and Matsumoto's advance sharing scheme for quantum shares significantly enhances the practicality and security of Quantum Secret Sharing. By integrating stabilizer codes and EAQECCs, their approach allows for efficient and secure distribution of quantum secrets, even in the presence of noise and partial information loss. This development represents a crucial step towards the broader adoption of quantum cryptographic techniques, promising more secure and flexible methods for protecting sensitive information in the quantum era.

4.3.4.1 Implementation Using Python `Pandas` and `Cirq`

This scheme is implemented using entangled pairs (Bell States) and distributing with Python `Pandas` and `Cirq`. The code is presented in Listing 4.7 and, while the result is shown in Figs. 4.16, 4.17, and 4.18.

The scheme leverages stabilizer codes and ramp quantum secret sharing for efficient and secure distribution of quantum secrets. This scheme allows for some shares to be distributed in advance, improving flexibility in secret sharing operations. The scheme from [7] involves using entanglement-assisted quantum error-correcting codes to distribute shares of a secret. This scheme leverages maximally entangled pairs and stabilizer codes to ensure the security and recoverability of the secret.

Below is a simplified implementation of this scheme using `Cirq` library. This implementation will focus on creating maximally entangled pairs (Bell states) and distributing them to the participants.

```
import cirq
import matplotlib.pyplot as plt
import numpy as np
# Create qubits for Alice, Bob, Charlie, and an ancilla qubit for entanglement
alice_qubit = cirq.NamedQubit('alice')
bob_qubit = cirq.NamedQubit('bob')
charlie_qubit = cirq.NamedQubit('charlie')
ancilla_qubit = cirq.NamedQubit('ancilla')
# Define a circuit to create Bell states and perform the protocol
```

```
def create_bell_pair(qubit1, qubit2):
circuit = cirq.Circuit()
# Create Bell state: (|00> + |11>)/sqrt(2)
circuit.append(cirq.H(qubit1)) # Apply Hadamard gate to the first qubit
circuit.append(cirq.CNOT(qubit1, qubit2)) # CNOT gate with the first qubit as control and the second as target
return circuit
def distribute_shares():
circuit = cirq.Circuit()
# Create Bell pairs: (|00> + |11>)/sqrt(2)
circuit += create_bell_pair(alice_qubit, ancilla_qubit)
circuit += create_bell_pair(bob_qubit, ancilla_qubit)
circuit += create_bell_pair(charlie_qubit, ancilla_qubit)
return circuit
def measure_shares():
circuit = cirq.Circuit()
# Measure in the Z basis
circuit.append(cirq.measure(alice_qubit, key='alice'))
circuit.append(cirq.measure(bob_qubit, key='bob'))
circuit.append(cirq.measure(charlie_qubit, key='charlie'))
circuit.append(cirq.measure(ancilla_qubit, key='ancilla'))
return circuit
# Create the entanglement distribution circuit
distribution_circuit = distribute_shares()
print("Distribution Circuit:")
print(distribution_circuit)
# Measure the shares
measurement_circuit = measure_shares()
print("\nMeasurement Circuit:")
print(measurement_circuit)
# Combine the circuits
full_circuit = distribution_circuit + measurement_circuit
print("\nFull Circuit:")
print(full_circuit)
# Simulate the circuit
simulator = cirq.Simulator()
result = simulator.run(full_circuit, repetitions=1000)
print("\nSimulation Results:")
print(result)
# Get the results
results = result.data
# Plot the histogram
results_str = results.apply(lambda row: f"{row['alice']}{row['bob']}{row['charlie']}{row['ancilla']}", axis=1)
histogram = results_str.value_counts()
# Plot histogram
plt.figure(figsize=(10, 6))
plt.bar(histogram.index, histogram.values, color='blue')
plt.xlabel('Measurement Outcome')
plt.ylabel('Frequency')
plt.title('Measurement Outcome Histogram for Entanglement Shares')
plt.show()
```

Listing 4.7 Advance Sharing of Quantum Shares for Quantum Secrets using Entangled Pairs (Bell States) and Distributing—with Python Pandas and CIRQ framework

4.3.4.2 Implementation Using Python `Numpy` and `Cirq`

The second approach of implementation is based on entangled pairs (Bell States) and distributing with Python `Numpy` and `Cirq`. Listing 4.8 we have the implementation of the example from Sect. 4.3.4.1. The output can be seen in Figs. 4.19, 4.20, and 4.21.

```
import cirq
import matplotlib.pyplot as plt
import numpy as np
# Create qubits for Alice, Bob, Charlie, and an ancilla qubit for entanglement
alice_qubit = cirq.NamedQubit('alice')
bob_qubit = cirq.NamedQubit('bob')
charlie_qubit = cirq.NamedQubit('charlie')
```

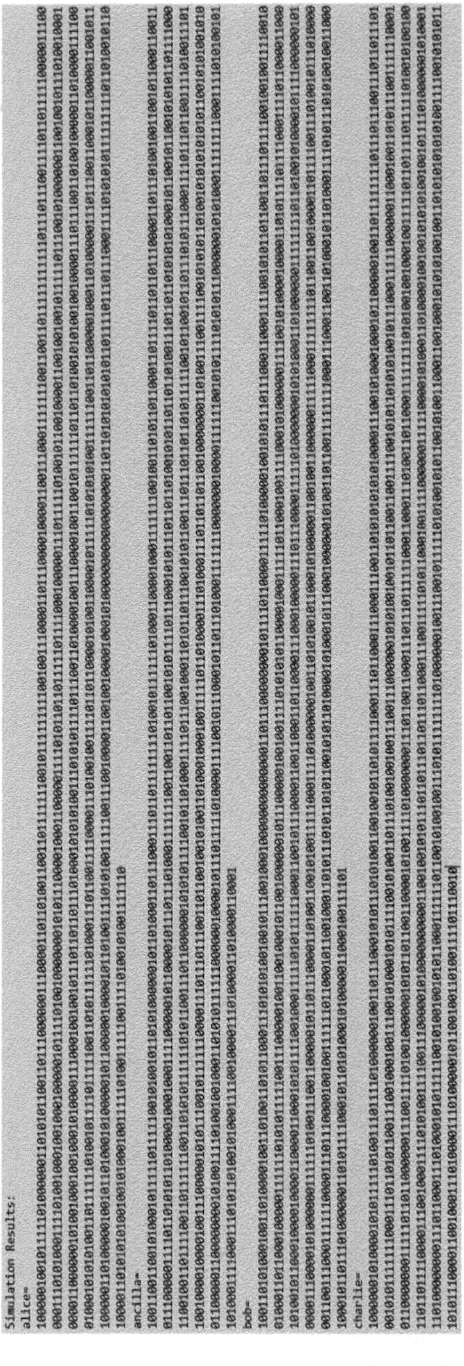

Fig. 4.16 Simulation results

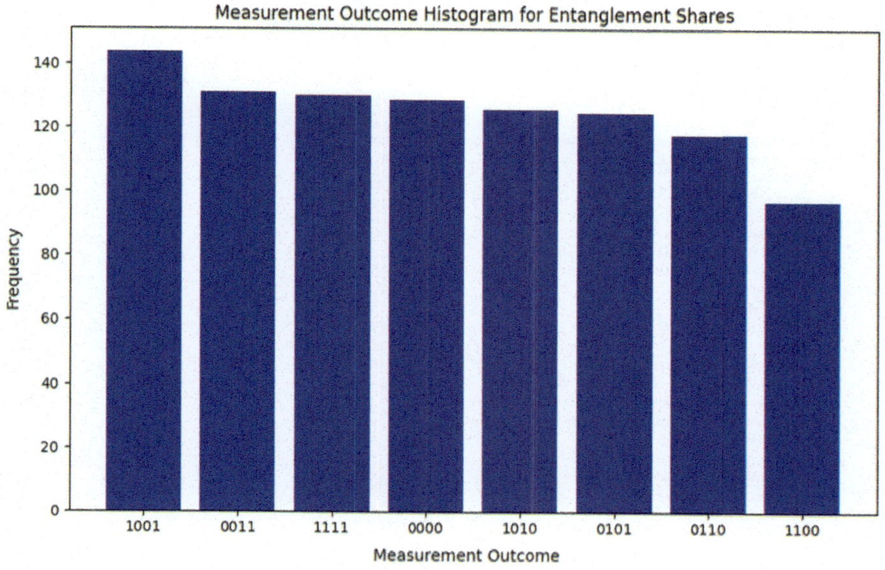

Fig. 4.17 Measurement of the outcome histogram for entanglement shares

Fig. 4.18 Distribution circuit, measurement circuit, and full circuit

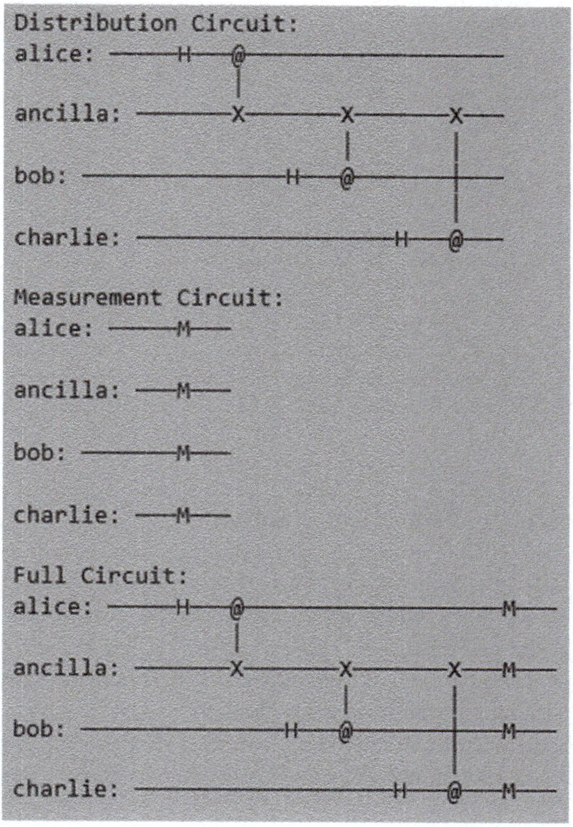

```
ancilla_qubit = cirq.NamedQubit('ancilla')
# Define a circuit to create Bell states and perform the protocol
def create_bell_pair(qubit1, qubit2):
circuit = cirq.Circuit()
# Create Bell state: (|00> + |11>)/sqrt(2)
circuit.append(cirq.H(qubit1)) # Apply Hadamard gate to the first qubit
circuit.append(cirq.CNOT(qubit1, qubit2)) # CNOT gate with the first qubit as control and the second as target
return circuit
def distribute_shares():
circuit = cirq.Circuit()
# Create Bell pairs: (|00> + |11>)/sqrt(2)
circuit += create_bell_pair(alice_qubit, ancilla_qubit)
circuit += create_bell_pair(bob_qubit, ancilla_qubit)
circuit += create_bell_pair(charlie_qubit, ancilla_qubit)
return circuit
def measure_shares():
circuit = cirq.Circuit()
# Measure in the Z basis
circuit.append(cirq.measure(alice_qubit, key='alice'))
circuit.append(cirq.measure(bob_qubit, key='bob'))
circuit.append(cirq.measure(charlie_qubit, key='charlie'))
circuit.append(cirq.measure(ancilla_qubit, key='ancilla'))
return circuit
# Create the entanglement distribution circuit
distribution_circuit = distribute_shares()
print("Distribution Circuit:")
print(distribution_circuit)
# Measure the shares
measurement_circuit = measure_shares()
print("\nMeasurement Circuit:")
print(measurement_circuit)
# Combine the circuits
full_circuit = distribution_circuit + measurement_circuit
print("\nFull Circuit:")
print(full_circuit)
# Simulate the circuit
simulator = cirq.Simulator()
result = simulator.run(full_circuit, repetitions=1000)
print("\nSimulation Results:")
print(result)
# Get the results
results = result.data
# Plot the histogram
results_str = results.apply(lambda row: f"{row['alice']}{row['bob']}{row['charlie']}{row['ancilla']}", axis=1)
histogram = results_str.value_counts()
# Plot histogram
plt.figure(figsize=(10, 6))
plt.bar(histogram.index, histogram.values, color='blue')
plt.xlabel('Measurement Outcome')
plt.ylabel('Frequency')
plt.title('Measurement Outcome Histogram for Entanglement Shares')
plt.show()
```

Listing 4.8 Advance Sharing of Quantum Shares for Quantum Secrets using Entangled Pairs (Bell States) and Distributing—with Python Numpy library and CIRQ framework

4.3.4.3 Performance Analysis

The given code provided in Listing 4.8 implements a complex quantum secret sharing mechanism, which distributes quantum shares of a secret among participants by utilizing Bell states. This performance analysis assesses the success of the building and simulation of the quantum circuit created for this purpose, while also highlighting potential areas for improvement.

The code initiates the creation of qubits for Alice, Bob, and Charlie, as well as an ancilla qubit for the purpose of entanglement. The primary concept is to generate

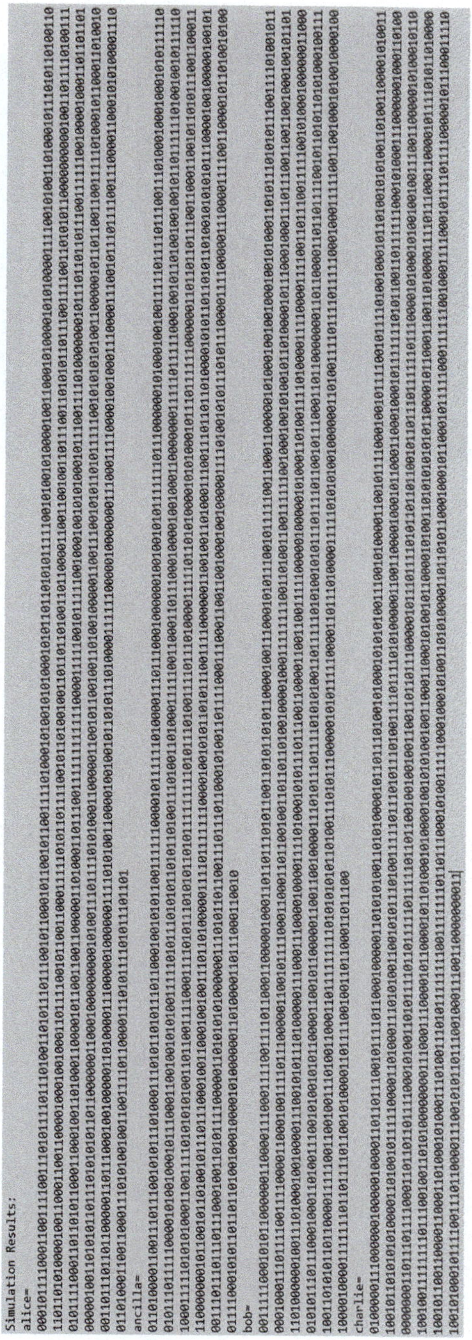

Fig. 4.19 Simulation results

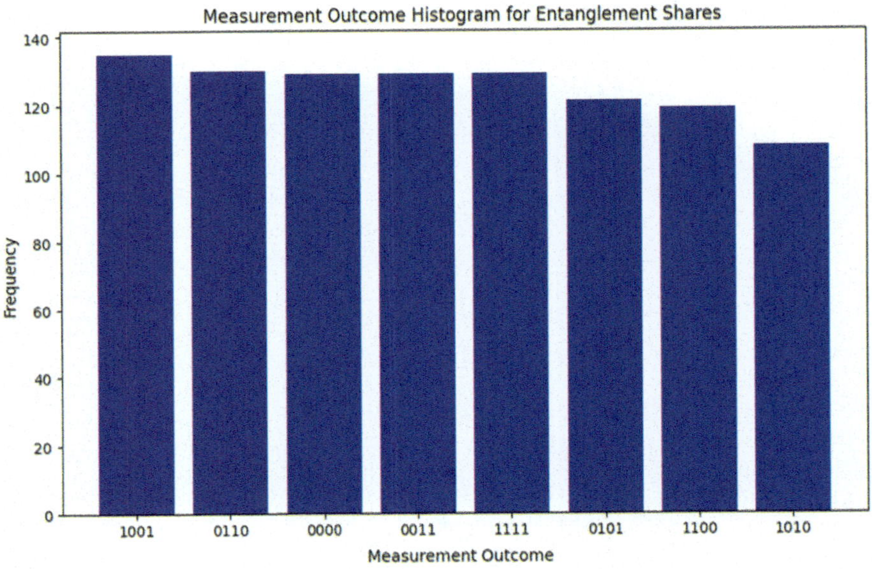

Fig. 4.20 Measurement of the outcome histogram for entanglement shares

Fig. 4.21 Distribution circuit, measurement circuit, and full circuit

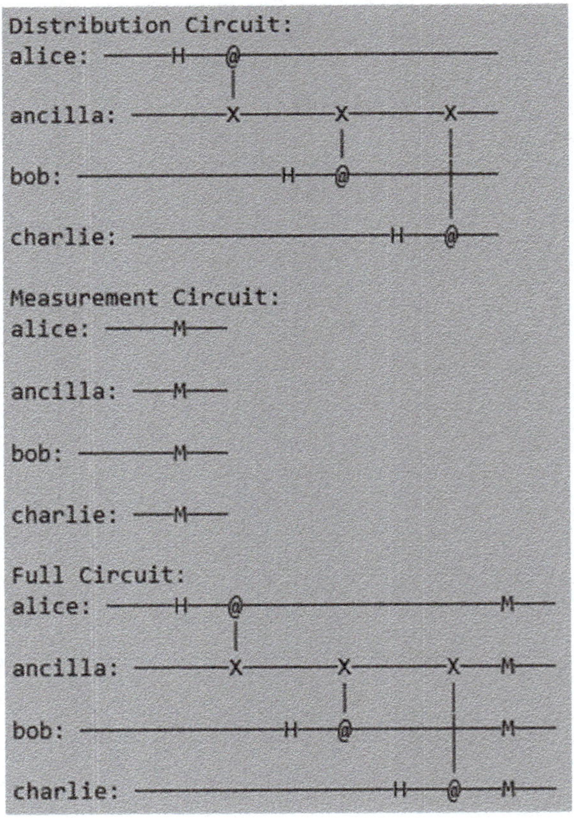

Bell states and allocate the portions of a quantum secret among the participants by utilizing these entangled states. Bell states refer to two-qubit states that are maximally entangled, such as $(|00\rangle + |11\rangle)/\sqrt{2}$. These states play an important role in numerous quantum communication protocols.

The *create_bell_pair* function is defined to establish a Bell state between two qubits. The first qubit is subjected to a Hadamard gate, resulting in a superposition. This is followed by a CNOT gate, where the first qubit acts as the control and the second qubit acts as the target. By performing this series of actions, the qubits become entangled, leading to the creation of the Bell state. The circuit depth required to create a Bell pair is low, consisting of only two gates. This characteristic is helpful as it preserve coherence in quantum states.

Afterwards, the *distribute_shares* function builds a circuit to generate and allocate the Bell pairs among the participants. This function invokes the *create_bell_pair* operation to entangle the qubit of each participant with the *ancilla qubit*. The circuit that is produced consists of three Bell pairs, with each pair entangling one participant (Alice, Bob, and Charlie) with the ancilla qubit. This configuration guarantees that the participants possess an entangled state, which is essential for the quantum secret sharing protocol.

After the shares are distributed, the *measure_shares* function creates a circuit to measure the states of the qubits in the Z basis. Every qubit, including the ancilla, is measured separately, and the outcomes are gathered. Next, the entanglement distribution and measurement circuits are simulated together as a combined circuit.

The complete circuit is simulated using `Cirq`'s simulator, with the circuit being executed 1000 times to collect statistical data on the measurement results. This level of repetition guarantees a strong dataset that accurately captures the probabilistic characteristics of quantum experiments. The efficacy of this simulation is heavily contingent upon the computational resources at hand. For a circuit with a limited number of qubits, usually four, the simulation should be relatively fast, usually completed within seconds.

The simulation findings are analyzed to develop a thorough comprehension of the measurement outcomes. The outcome of each measurement is translated into a string that represents the cumulative results for each qubit. A histogram is then created to visually display the frequency of distinct measurement outcomes. The histogram accurately shows the distribution of entangled states among the participants, offering valuable insights into the behavior of the quantum system.

Different enhancements can be explored for this implementation to increase its performance. Further optimization of the circuit depth could be achieved through the exploration of additional entanglement techniques, but the existing methodology is already efficient in the creation and measurement of Bell states. Scalability is an important factor, as well. As the number of players in the system increases, improvements in gate operations and parallelism can boost performance. In addition, it may be required to employ quantum error correction methods in order to preserve the accuracy of the quantum states.

To achieve extended simulations, utilizing cloud-based quantum computing resources, such as Google's Quantum Computing Service, can provide expedited outcomes. Augmenting the number of iterations can enhance statistical precision, albeit at the expense of increased simulation duration. Hence, it is crucial to maintain a balance between the requirement for precision and the computational resources that are now accessible.

To summarize, the given code successfully shows the fundamentals of a complex quantum secret sharing system, which distributes quantum shares by utilizing Bell states. The performance is adequate for simulations on a small scale. Nevertheless, when dealing with larger and more intricate implementations, it will be mandatory to make optimizations in both the quantum circuit and the simulation process in order to uphold efficiency and accuracy. This analysis emphasizes the promise and difficulties of using quantum algorithms for secure information exchange and emphasizes the significance of ongoing progress in quantum computing approaches.

4.3.5 Computational Quantum Secret Sharing

Quantum secret sharing allows a dealer to spread a secret quantum state among a group of parties. It guarantees that only specific permitted subsets of the group may reconstruct the secret, while unauthorized subsets are unable to get any information about it. In the past, QSS schemes primarily emphasized achieving complete security, which required shares that grew exponentially in size, even for access structures that could be computed by circuits of polynomial size. In contrast to traditional secret sharing, this approach differs significantly by allowing for computationally secure schemes that have polynomial-time complexity and shorter shares, all based on normal hardness assumptions.

The research conducted by Alper Cakan, Vipul Goyal, Chen-Da Liu-Zhang, and João Ribeiro [8] and proposed in 2023 on Computational Quantum Secret Sharing (CQSS) is innovative as it investigates the advantages of computational security in quantum environments, which is comparable to the principles observed in classical encryption. Their study presents a compiler that builds computationally safe QSS schemes based on common hardness assumptions, resulting in a notable reduction in share size compared to earlier approaches. An important breakthrough highlighted in their work is the capability to create QSS schemes with significantly smaller quantum shares compared to the size of the secret. This achievement is not possible to reach with absolute security. This advancement has a significant effect on access structures that previously necessitated exponential share sizes. Their method not only improves the effectiveness of QSS schemes but also expands their suitability to more intricate and practical situations. In addition, their compiler has the capability to optimize the size of the shared information in the context of information theory. They enhanced the share size of flawless QSS schemes to $1.5^{n+o(n)}$ for a significant range of nn-party access arrangements, which aligns with the most advanced findings in the traditional context. This optimization proves the capability of CQSS to attain efficiency close to that of classical systems, while yet ensuring strong quantum security. In addition, the authors examine the category of access structures that can be effectively executed when the QSS scheme has access to many duplicates of the secret. This encompasses operations within the complexity classes PP and NP, providing a thorough foundation for comprehending the computational complexities linked to different QSS schemes. Overall, the research conducted by Cakan, Goyal, Liu-Zhang, and Ribeiro makes substantial progress in the realm of quantum cryptography by utilizing computational security to enhance the efficiency and feasibility of QSS schemes. Their innovations facilitate the broader implementation of quantum cryptography algorithms, guaranteeing secure and efficient information exchange in quantum networks. To obtain a comprehensive analysis of their techniques and findings, the reader may consult their complete research paper from the Cryptology ePrint Archive [8].

4.3.5.1 Implementation

The scheme introduces the concept of computationally-secure quantum secret sharing, focusing on reducing the share size under standard computational hardness assumptions. This is a more advanced concept, but we can create a simplified implementation using `Cirq` framework to illustrate the principles involved as it can be seen in Listing 4.9.

```
import cirq
import matplotlib.pyplot as plt
import pandas as pd

# Create qubits for Alice, Bob, Charlie, and an ancilla qubit for entanglement
alice_qubit = cirq.NamedQubit('alice')
bob_qubit = cirq.NamedQubit('bob')
charlie_qubit = cirq.NamedQubit('charlie')
ancilla_qubit = cirq.NamedQubit('ancilla')

# Define a circuit to create entangled states and perform the protocol
def create_entanglement():
circuit = cirq.Circuit()

# Create entanglement: (|0000> + |1111>)/sqrt(2)
circuit.append(cirq.H(alice_qubit)) # Apply Hadamard gate to Alice's qubit
circuit.append(cirq.CNOT(alice_qubit, bob_qubit)) # CNOT gate with Alice as control and Bob as target
circuit.append(cirq.CNOT(bob_qubit, charlie_qubit)) # CNOT gate with Bob as control and Charlie as target
circuit.append(cirq.CNOT(charlie_qubit, ancilla_qubit)) # CNOT gate with Charlie as control
# and Ancilla as target

return circuit

def measure_entanglement():
circuit = cirq.Circuit()

# Measure in the Z basis
circuit.append(cirq.measure(alice_qubit, key='alice'))
circuit.append(cirq.measure(bob_qubit, key='bob'))
circuit.append(cirq.measure(charlie_qubit, key='charlie'))
circuit.append(cirq.measure(ancilla_qubit, key='ancilla'))

return circuit

# Create the entanglement distribution circuit
entanglement_circuit = create_entanglement()
print("Entanglement Circuit:")
print(entanglement_circuit)
```

```python

    # Measure the entanglement
    measurement_circuit = measure_entanglement()
    print("\nMeasurement Circuit:")
    print(measurement_circuit)

    # Combine the circuits
    full_circuit = entanglement_circuit + measurement_circuit
    print("\nFull Circuit:")
    print(full_circuit)

    # Simulate the circuit
    simulator = cirq.Simulator()
    result = simulator.run(full_circuit, repetitions=1000)
    print("\nSimulation Results:")
    print(result)

    # Get the results
    results = result.data

    # Plot the histogram
    results_str = results.apply(lambda row: f"{row['alice']}{row['bob']}{row['charlie']}{row['ancilla']}", axis=1)
    histogram = results_str.value_counts()

    # Plot histogram
    plt.figure(figsize=(10, 6))
    plt.bar(histogram.index, histogram.values, color='blue')
    plt.xlabel('Measurement Outcome')
    plt.ylabel('Frequency')
    plt.title('Measurement Outcome Histogram for Entanglement Shares')
    plt.show()
```

Listing 4.9 Implementation of computational quantum secret sharing

4.3.5.2 Performance Analysis

The given code from Listing 4.9 utilizes entangled states to create a computational quantum secret sharing (CQSS) mechanism. This performance analysis assesses the success of the building and simulation of the quantum circuit built for this purpose, as well as identifies potential areas for improvement (Fig. 4.22).

The code initiates the creation of qubits for Alice, Bob, and Charlie, as well as an ancilla qubit for the purpose of entanglement. The primary objective is to generate an entangled state among these qubits and thereafter measure the entangled states to distribute a quantum secret among the participants (Fig. 4.23).

The *create_entanglement* function generates a quantum circuit to generate an entangled state. This function applies a Hadamard gate on the qubit owned by Alice, resulting in the creation of a superposition state. Subsequently, it employs a sequence of CNOT gates to transmit this superposition throughout the remaining qubits, resulting in the formation of a maximally entangled state. More precisely, the circuit utilizes:

- An application of a Hadamard gate to the qubit owned by Alice,
- A CNOT gate is used with Alice's qubit as the control and Bob's qubit as the target,
- A CNOT gate with Bob's qubit as the control and Charlie's qubit as the target,
- A CNOT gate is applied to the qubits, with Charlie's qubit serving as the control qubit and the ancilla qubit as the target qubit.

Fig. 4.22 Simulation results

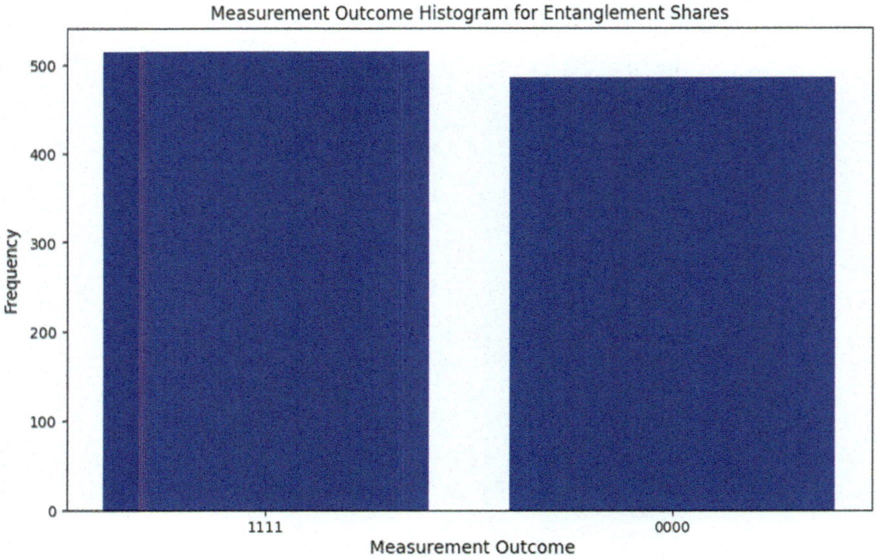

Fig. 4.23 Measurement of the outcome histogram for entanglement shares

The series of actions leads to the creation of the entangled state $(|0000\rangle + |1111\rangle)/\sqrt{2}$. The circuit depth required to create this entangled state is small, consisting of only four gates. This is beneficial for preserving coherence in quantum states (Fig. 4.24).

The *measure_entanglement* function establishes a circuit to measure the quantum states of the qubits in the Z basis. Every qubit, including the ancilla, is measured separately, and the outcomes are gathered. The complete circuit consists of the entanglement and measurement circuits combined, encompassing all the required activities for both generating and measuring the entangled state.

The complete circuit is simulated using `Cirq`'s simulator, with the circuit being executed 1000 times to collect statistical data on the measurement results. This level of repetition guarantees a strong dataset that accurately captures the probabilistic characteristics of quantum experiments. The effectiveness of this simulation is heavily reliant on the computational resources at hand. For a circuit with a limited number of qubits, typically four, the simulation should be relatively fast, usually completed within seconds.

The simulation results are analyzed to develop a full understanding of the measurement outcomes. The outcome of each measurement is translated into a string that represents the cumulative results for each qubit. A histogram is then created to visually display the frequency of distinct measurement outcomes. This histogram accurately depicts the distribution of the entangled states among the players, offering valuable insights into the behavior of the quantum system.

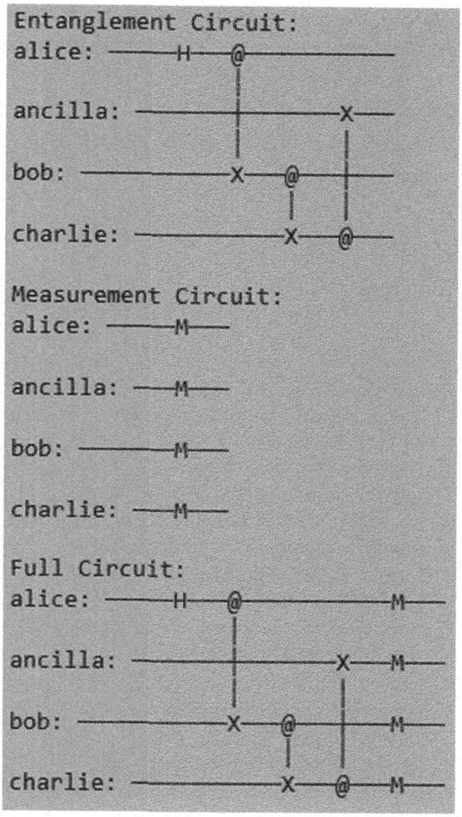

Fig. 4.24 Distribution circuit, measurement circuit, and full circuit

Various enhancements can be made for this implementation to increase its performance. Further optimization of the circuit depth could be achieved by the exploration of improved entanglement techniques. However, the existing methodology is already efficient in the creation and measurement of entangled states. Scalability is another important factor to consider. As the number of players in the system increases, improvements in gate operations and parallelism can boost performance. Furthermore, the use of quantum error correction techniques may be required in order to increase the accuracy and integrity of the quantum states.

To achieve extended simulations, utilizing cloud-based quantum computing resources, such as Google's Quantum Computing Service, can provide expedited outcomes. Augmenting the quantity of iterations can enhance the precision of statistical measurements, albeit at the expense of increased simulation duration. Hence, it is important to find a balance between the requirement for precision and the computational resources at hand.

Overall, the given code successfully proves the concepts of the Computational Quantum Secret Sharing method, which involves the distribution of quantum shares through the use of entangled states. The performance is adequate for simulations of a small scale. However, in the case of bigger and more intricate implementations, it will be imperative to make optimizations in both the quantum circuit and the simulation process in order to uphold efficiency and accuracy. This analysis emphasizes the possibilities and difficulties of using quantum algorithms for secure information exchange and emphasizes the significance of ongoing progress in quantum computing approaches.

References

1. Boddu, N.G., et al.: Split-State Non-Malleable Codes and Secret Sharing Schemes for Quantum Messages. 2024/335, 2024. Cryptology ePrint Archive, Cryptology ePrint Archive (eprint.iacr.org), https://eprint.iacr.org/2024/335
2. Li, G.-D., Cheng, W.-C., Wang, Q.-L., Cheng, L., Mao, Y., Jia, H.-Y.: Quantum Secret Sharing Enhanced: Utilizing W States for Anonymous and Secure Communication. Ar5iv, https://ar5iv.labs.arxiv.org/html/2402.02413. Accessed 11 June 2024
3. Masumori, S., Matsumoto, R.: Advance Sharing with Ogawa et al.'s Ramp Quantum Secret Sharing Scheme. Ar5iv, https://ar5iv.labs.arxiv.org/html/2404.15646. Accessed 11 June 2024
4. Shamir, A.: How to share a secret. Commun. ACM **22**, 11, 612–613 (1979). https://doi.org/10.1145/359168.359176
5. Blakley, G.R.: Safeguarding cryptographic keys. In: 1979 International Workshop on Managing Requirements Knowledge (MARK) (1899), pp. 313–318
6. Ogawa, T., Sasaki, A., Iwamoto, M., Yamamoto, H.: Quantum secret sharing schemes and reversibility of quantum operations. Phys. Rev. A **72** (2005). https://doi.org/10.1103/PhysRevA.72.032318
7. Shibata, M., Matsumoto, R.: Advance sharing of quantum shares for quantum secrets. Ar5iv, https://ar5iv.labs.arxiv.org/html/2302.14448. Accessed 11 June 2024
8. Cakan, A., et al.: Computational Quantum Secret Sharing. 2023/613, 2023. Cryptology ePrint Archive, Cryptology ePrint Archive (eprint.iacr.org), https://eprint.iacr.org/2023/613

Chapter 5
Quantum Computing Applications

Quantum computing is a rapidly evolving field that has the potential to transform many aspects of science and industry.

Recent advances in quantum computing, such as Google establishing quantum dominance,[1] have brought interest to this domain. However, there is limited knowledge about how quantum computing will affect enterprises. Our research points to the following categories of quantum computing use cases and real-world applications for 2024: Optimization research, cryptography/espionage, industry-specific applications; quantum computing is projected to have the greatest impact in finance, materials research, and healthcare.

Companies such as Google, IBM, and Microsoft are taking the first moves toward creating quantum computing platforms. Many other companies hope to use quantum computing to spearhead pioneering research in their fields of specialization. For example, Bosch[2] plans to use quantum simulations to develop new energy storage and useful materials.

According to IBM, the most important use cases of the quantum computing will evolve following three maturity horizons (see Fig. 5.1[3]), such as:

- 1st horizon: identifying the most important applications and fields for the next years.
- $2^{nd} horizon$: this will take place as a stable pillar but will not be optimally for the current working quantum computers.
- 3rd horizon: going beyond the 15 years.

[1] https://www.nature.com/articles/s41586-019-1666-5.

[2] https://www.bosch.com/research/fields-of-innovation/quantum-technologies/.

[3] https://research.aimultiple.com/quantum-computing-applications/.

M. I. Mihailescu et al., *Applied Quantum Computing and Cryptography*,
Studies in Big Data 167, https://doi.org/10.1007/978-3-031-77145-3_5

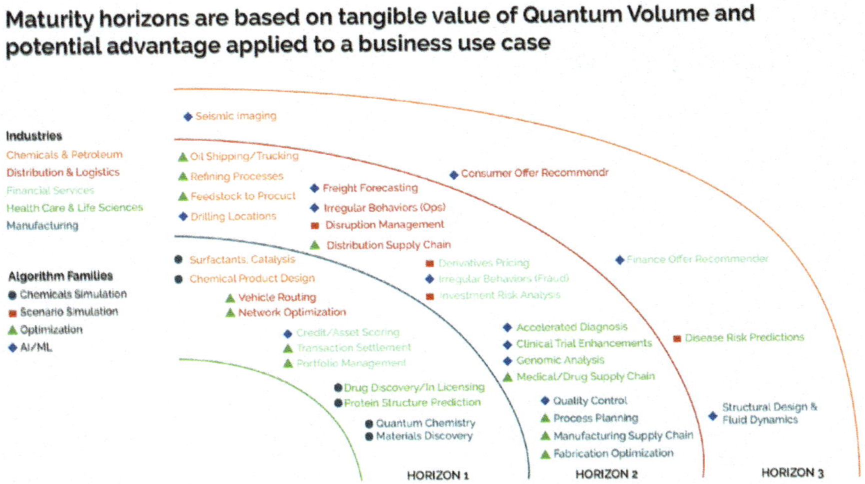

Fig. 5.1 Maturity horizons for the future of quantum computing application

Below we have some major uses of quantum computing:

- **Cryptography**
  - *Quantum Key Distribution (QKD)*. Enables safe communication by distributing encryption keys with theoretically unbreakable security.
  - *Breaking classical cryptography*. Quantum computers have the capacity to solve problems such as integer factorization (used in RSA encryption) exponentially quicker than classical computers, posing a danger to existing encryption systems.
- **Drug discovery and material science**
  - *Molecular Simulation*. Quantum computers can simulate complicated molecular structures and interactions more efficiently than classical computers, which speeds up the development of new medications and materials.
  - *Optimization of chemical reactions*. Assists in determining the ideal paths for chemical reactions, resulting in more efficient and ecologically friendly operations.
- **Optimization cases and problems**
  - *Supply chain management*. Quantum algorithms can optimize logistics and supply chain operations, lowering costs while increasing efficiency.
  - *Financial portfolio optimization*. Can tackle complicated optimization issues in finance, such as portfolio management and risk assessment, more efficiently.

- **Machine Learning and Artificial Intelligence**
  - *Quantum Machine Learning*. Quantum computers can analyze large volumes of data and execute difficult computations at rapid rates, potentially improving AI and ML by allowing for faster model training and more accurate predictions.
  - *Pattern Recognition*. Improved ability to recognize patterns within massive datasets, which can be applied in a variety of industries like healthcare, finance, and cybersecurity.
- **Computational Biology**
  - *Protein folding*. Predicting the three-dimensional structure of proteins, which is critical for understanding diseases and finding new treatments.
  - *Genomic analysis* is the process of analyzing genetic data faster in order to find mutations and better understand genetic illnesses.
- **Energy sector**
  - *Optimization of energy networks*. Improving energy management and distribution in smart networks to increase efficiency and reduce losses.
  - *Simulating and identifying novel materials* for use in more efficient batteries and energy storage systems.
- **Climate changes and modeling**
  - Enhanced Simulations: Provides reliable climate models, which aid in the forecast and mitigation of climate change effects.
- **Airspace, aerospace and defense**
  - *Flying path optimization* increases flying efficiency and safety by optimizing routes and fuel usage.
  - *Cryptographic security* ensures safe communication in defense using quantum-resistant encryption technologies.
- *Fundamental science*
  - *High-energy physics*. Extends classical computational capabilities by simulating quantum field theories and particle interactions.
  - *Astrophysics* involves analyzing and simulating complex astronomical events.
- *Algorithm development*
  - *Quantum algorithms*. Creates novel algorithms that use quantum mechanics, such as Grover's and Shor's algorithms, to tackle certain problems more effectively.

Optimisation challenges exist in all industries and business processes, and some of them take too long to address optimally using conventional computers. Two quantum techniques can be utilized to overcome these difficulties.

Quantum annealing[4] is an optimization heuristic that is projected to outperform traditional computers in some optimization tasks. Quantum annealing can be implemented on specialized quantum annealers, which are far simpler to create than a universal quantum computer. Currently, such computers are available; nevertheless, their superiority over traditional computers has yet to be established. Cheaper digital annealers replicate quantum annealers using classical computation, providing cost-effective alternatives.

Universal quantum computers are capable of tackling any form of computing issue. However, they will take longer to become commercially available because additional study is required to improve their reliability.

Some optimization problems from many industries, where some corporations are investing in quantum computing rather than suboptimal heuristics, are given below:

1. *Automotive*. Optimizing big autonomous fleets–Volkswagen[5] and Google collaborated to build autonomous vehicles using quantum computing.
2. *Energy*
   - *Utilization prediction*. Since 2020, the Dubai Electricity and Water Authority (DEWA) has collaborated with Microsoft[6] to use quantum computing for energy optimization.
   - *Grid optimization*. The United States Department of Energy (DOE)[7] has established two labs dedicated to integrating quantum computing into grid optimization.
   - *Weather forecasting*. IBM[8] employs quantum computing for precise weather forecasting, and it currently serves practically all consumer tech providers, including Apple, Amazon, Google, and Facebook.
3. *Finance*
   - Automated trading, such as anticipating financial markets.
   - *Risk analysis*. In 2020, Caixa Bank[9] launched a pilot initiative to test quantum computing for risk classification algorithms.
   - *Portfolio optimization*. KPMG[10] conducted a portfolio management benchmarking test in 2021 and discovered that quantum computing outperformed all other benchmarking methods for near-term investments.

[4] https://research.aimultiple.com/quantum-annealing/.

[5] https://www.vw.com/en/newsroom/future-of-mobility/quantum-computing.html.

[6] https://tinyurl.com/4yk7cdm5.

[7] https://www.energy.gov/science/doe-explainsquantum-computing.

[8] https://tinyurl.com/yc6pn8yj.

[9] https://tinyurl.com/4cpbxjw2.

[10] https://tinyurl.com/w8sk3h3z.

- *Fraud detection.* In 2022, PayPal[11] will collaborate with IBM to deploy quantum computing in fraud detection.

4. *Insurance*

- *Valuation of instruments and premiums in complex cases.* Anthem,[12] a health insurance firm, plans to deploy IBM's quantum computing solution for data-intensive tasks including recognizing health irregularities.

5. *Logistics*

- *Route and traffic optimization.* DWave,[13] a QC business, collaborated with Toyota to conduct research demonstrating that using QC for traffic prediction and optimization outperforms conventional alternatives.
- *Supply chain and inventory optimization.* While QC application is not yet a well-established solution for supply chain optimization, major corporations like Coca-Cola Japan[14] have begun large-scale experiments.

6. *Manufacturing*

- *Design optimization (e.g. batteries, chips, automobiles, etc.).* Daimler[15] is collaborating with IBM's quantum computing technology to create longer-lasting semiconductors.

7. *Pharma*

- *Drug interaction prediction.* A Massachusetts-based biotechnology company has collaborated[16] with 1QBit to use quantum computing for molecular comparison in drug development.
- *Personalized therapy using genomics.* Cambridge Quantum,[17] a research partner of Roche and Crownbio, is developing a quantum computer technology for analyzing genetic data for cancer treatment.

8. *Technology/software*

- *Machine learning.* Quantum computers may dramatically enhance machine learning operations.[18]

[11] https://tinyurl.com/y9cdud38.
[12] https://www.techrepublic.com/topic/big-data/.
[13] https://www.dwavesys.com/media/b4lllchm/dwave_toyota_case_story_v4.pdf.
[14] https://tinyurl.com/dr84ter5.
[15] https://www.ibm.com/quantum/blog/next-gen-lithium-sulfur-batteries.
[16] https://cen.acs.org/business/informatics/Lets-talk-quantum-computing-drug/98/i35.
[17] https://tinyurl.com/3v555c8x.
[18] https://research.aimultiple.com/quantum-ai/.

While many applications of quantum computing are still in the experimental stage, the potential for substantial breakthroughs is at a high rate. As technology progresses, practical applications will become possible, transforming different fields and encouraging innovation. With its cutting-edge technology, quantum computing has the potential to transform many other industries, including materials science and encryption.

Modern encryption methods. Among the many potential uses of quantum computing, quantum cryptography stands out. The enormous processing power of quantum computers makes traditional cryptography algorithms like RSA and ECC simply breakable. The possibility for quantum-mechanically-based, impenetrable encryption is demonstrated through the analysis of quantum key distribution. Two parties can safely exchange encryption keys using QKD, which makes use of phenomena like entanglement and superposition, making any effort at eavesdropping immediately evident. In a world where quantum computing is becoming more and more of a concern, this research is vital for protecting communications in many different industries, such as banking, government, and critical infrastructure.

In addition, quantum cryptography may have contributions in searchable and homomorphic encryption. To keep data private and secure, these encryption methods permit calculations on encrypted data without decrypting it. Current homomorphic encryption methods based on lattices are considered to be secure against quantum attacks by combining quantum-resistant approaches. This is important for applications in cloud computing and large data.

Dynamic Algorithms. Mastering the capabilities of quantum computers includes creating and refining quantum algorithms. For example, in these chapters, we have seen how important quantum algorithms, including Shor's and Grover's, affect classical cryptosystems. For instance, conventional methods are outpaced by Shor's algorithm when it comes to factorizing huge integers, which directly undermines RSA and other cryptographic systems.

Finding these weaknesses, we can highlight how critical it is to provide cryptographic solutions that are resistant to quantum computing. The cryptography community needs to be proactive in responding to these threats by adapting and inventing new ways to secure critical data even in the quantum future. To ensure that cryptography continues to advance and push the limits of what can be achieved with quantum-resistant approaches, it is essential to take this proactive approach.

Computer-Aided Design. Quantum computing's usefulness in computational modeling and simulation extends well beyond the realm of cryptography. The wider influence of computer modeling is demonstrated by research contributions to computational approaches for studies in the natural sciences, including bungee leaping and rocket motion. Simulating quantum systems, which are fundamentally complicated and necessitate high-end computational methods, is a perfect fit for these methodologies.

Research on material sciences is another example of the possibilities offered by quantum simulations. Advancements in material design and discovery have been made possible by the quantum computer's ability to simulate atomic and molecule interactions with an unprecedented level of accuracy. Because quantum simulations are so much better at predicting molecular behavior and interactions than classical simulations, this capacity is especially important in areas like drug discovery.

Chapter 6
Future Directions and Open Challenges

Quantum computing is on the verge of changing various technological domains, and cryptography is expected to be significantly affected. With the continuous progress of quantum computing, there is an increasing need for computing solutions that are secure against quantum attacks and capable of handling the difficult problems brought about by this new era.

An area that needs concentrated investigation is the advancement of quantum-resistant algorithms. Present cryptography techniques, such as RSA and ECC, are susceptible to the computational capabilities of quantum computers. Therefore, it is crucial to investigate and create new mathematical frameworks that can support algorithms that are resistant to quantum computing attacks. This includes the examination of lattice-based cryptography, hash-based cryptography, code-based cryptography, and multivariate polynomial cryptography. Furthermore, it is important to conduct research on hybrid cryptographic systems that include both classical and quantum-resistant algorithms in order to guarantee strong security while transitioning into the era of quantum computing.

Progress in quantum key distribution also offers a hopeful path forward. Although quantum key distribution provides encryption that is theoretically impossible to break, its existing implementations are limited by factors such as distance limitations and the rate at which keys may be generated. Subsequent investigations should focus on the development of novel and enhanced quantum key distribution protocols that effectively tackle these constraints, hence enhancing the practicality and widespread adoption of it. It is essential to include QKD into current communication infrastructures, such as internet and satellite systems, in order to facilitate wider acceptance. Thorough security evaluations of these protocols are required to detect and address potential weaknesses, guaranteeing their resistance to advancing quantum attacks.

The focus on post-quantum cryptography (PQC) standards is is also an important topic. By actively collaborating with international standardization organizations, we may actively participate in the advancement and establishment of PQC algorithms,

M. I. Mihailescu et al., *Applied Quantum Computing and Cryptography*,
Studies in Big Data 167, https://doi.org/10.1007/978-3-031-77145-3_6

guaranteeing that they fulfill the essential requirements for security and performance. Developing comprehensive implementation guidelines and optimal strategies for enterprises seeking to use PQC will ease the transition process and effectively reduce potential risks.

Developing network topologies that are resistant to quantum attacks will be essential for guaranteeing complete security in the era of quantum computing. Further investigations should prioritize the development of network topologies that integrate quantum-secure communication protocols. The development of methods to seamlessly integrate quantum and classical network components will enable a gradual transition to networks that are totally secure against quantum threats.

Quantum-secure computing is essential for ensuring the security of cloud computing. It is important to research and apply quantum-safe encryption methods specifically designed for cloud environments in order to ensure the security of data when it is stored, transmitted, and processed. The development and improvement of quantum-resistant secure multi-party computation protocols will allow for the execution of calculations that protect privacy in cloud environments. Integrating current cloud infrastructure to facilitate quantum-safe processes, such as secure key management and distribution techniques, will enhance the security of cloud-based applications.

The advancement of quantum cryptography hardware is also crucial. The development of hardware security modules (HSMs) and quantum random number generators (QRNGs) that are resistant to quantum attacks will ensure the availability of hardware that can support cryptographic operations that are secure against quantum threats. In addition, the development of tamper-resistant devices that can securely execute algorithms resistant to quantum attacks would provide protection against physical attacks, hence improving overall security.

Education and the development of the workforce are also crucial aspects of future progress. Creating extensive educational programs and curriculum centered on quantum computing and quantum-secure encryption will equip the future generations with the necessary expertise. Conducting workshops, seminars, and conferences will promote the sharing of knowledge and encourage cooperation among scholars, industry experts, and policymakers. Public awareness campaigns should inform both companies and individuals about the significance of adopting quantum-secure technologies and the necessary measures they may implement to get ready.

Collaborations involving other disciplines will also be essential. Engaging in partnerships with physicists, engineers, and other experts allows for a comprehensive approach to tackling the many obstacles of quantum-secure computing. Forming collaborations with sectors such as finance, healthcare, and government to test and improve quantum-secure systems in real-life environments will yield significant knowledge and tangible uses. International collaboration will facilitate the exchange of knowledge, resources, and optimal methods, addressing worldwide issues in quantum-secure computing.

Ultimately, the outlook for quantum-secure computing is filled with both promise and difficulty. Significant progress can be achieved by prioritizing the advancement of quantum-resistant algorithms, improvements in quantum key distribution

QKD, the establishment of post-quantum cryptography (PQC) standards, the development of quantum-secure network architectures, the implementation of quantum-safe cloud computing, the creation of quantum cryptographic hardware, investments in educational and workforce development, and fostering interdisciplinary research collaborations.

Chapter 7
Conclusion

Quantum computing represents a massive increase in computer capacity, leveraging quantum physics to do calculations at unprecedented rates. Unlike classical computers, which are confined by binary bits, quantum computers use qubits, which can exist in several states at once due to superposition. This enables them to process massive volumes of data and solve complicated issues significantly faster than traditional approaches.

Quantum computing has far-reaching consequences, with the potential to transform industries as diverse as encryption, drug development, materials research, and artificial intelligence. However, this unprecedented power offers a huge challenge to traditional encryption approaches, as quantum computers can break widely used encryption schemes through fast factorization and discrete logarithm calculations. This emphasizes the essential necessity for quantum cryptography.

Quantum key distribution is a breakthrough way to secure communication that allows two parties to generate a shared secret key using qubits. Eavesdropping efforts are instantly apparent because to quantum physics principles, providing data secrecy and integrity even in the face of quantum computing's tremendous powers.

In conclusion, while quantum computing has the potential to alter many fields and solve complicated issues, it also needs the development of better cryptographic technologies to protect information in this new era of computational power.

M. I. Mihailescu et al., *Applied Quantum Computing and Cryptography*,
Studies in Big Data 167, https://doi.org/10.1007/978-3-031-77145-3_7

Index

M. I. Mihailescu et al., *Applied Quantum Computing and Cryptography*, Studies in Big Data 167, https://doi.org/10.1007/978-3-031-77145-3

The manufacturer's authorised representative in the EU is Springer Nature Customer Service Centre GmbH, Europaplatz 3, 69115 Heidelberg, Germany. If you have any concerns regarding our products, please contact ProductSafety@springernature.com

Printed and bound by CPI Group (UK) Ltd, Croydon, CR0 4YY
30/09/2026
02225471-0016